T0280931

A Primer on Hardware Prefetching

Synthesis Lectures on Computer Architecture

Editor
Margaret Martonosi, *Princeton University*
Founding Editor Emeritus
Mark D. Hill, *University of Wisconsin, Madison*

Synthesis Lectures on Computer Architecture publishes 50- to 100-page publications on topics pertaining to the science and art of designing, analyzing, selecting and interconnecting hardware components to create computers that meet functional, performance and cost goals. The scope will largely follow the purview of premier computer architecture conferences, such as ISCA, HPCA, MICRO, and ASPLOS.

A Primer on Hardware Prefetching
Babak Falsafi and Thomas F. Wenisch
June 2014

On-Chip Photonic Interconnects: A Computer Architect's Perspective
Christopher J. Nitta, Matthew K. Farrens, Venkatesh Akella
October 2013

Optimization and Mathematical Modeling in Computer Architecture
Tony Nowatzki, Michael Ferris, Karthikeyan Sankaralingam, Cristian Estan, Nilay Vaish, David Wood
September 2013

Security Basics for Computer Architects
Ruby B. Lee
September 2013

The Datacenter as a Computer: An Introduction to the Design of Warehouse-Scale Machines, Second edition
Luiz André Barroso, Jimmy Clidaras, Urs Hölzle
July 2013

Shared-Memory Synchronization
Michael L. Scott
June 2013

Resilient Architecture Design for Voltage Variation
Vijay Janapa Reddi, Meeta Sharma Gupta
June 2013

Multithreading Architecture
Mario Nemirovsky, Dean M. Tullsen
January 2013

Performance Analysis and Tuning for General Purpose Graphics Processing Units (GPGPU)
Hyesoon Kim, Richard Vuduc, Sara Baghsorkhi, Jee Choi, Wen-mei Hwu
November 2012

Automatic Parallelization: An Overview of Fundamental Compiler Techniques
Samuel P. Midkiff
January 2012

Phase Change Memory: From Devices to Systems
Moinuddin K. Qureshi, Sudhanva Gurumurthi, Bipin Rajendran
November 2011

Multi-Core Cache Hierarchies
Rajeev Balasubramonian, Norman P. Jouppi, Naveen Muralimanohar
November 2011

A Primer on Memory Consistency and Cache Coherence
Daniel J. Sorin, Mark D. Hill, David A. Wood
November 2011

Dynamic Binary Modification: Tools, Techniques, and Applications
Kim Hazelwood
March 2011

Quantum Computing for Computer Architects, Second Edition
Tzvetan S. Metodi, Arvin I. Faruque, Frederic T. Chong
March 2011

High Performance Datacenter Networks: Architectures, Algorithms, and Opportunities
Dennis Abts, John Kim
March 2011

Processor Microarchitecture: An Implementation Perspective
Antonio González, Fernando Latorre, Grigorios Magklis
December 2010

Transactional Memory, 2nd edition
Tim Harris, James Larus, Ravi Rajwar
December 2010

Computer Architecture Performance Evaluation Methods
Lieven Eeckhout
December 2010

Introduction to Reconfigurable Supercomputing
Marco Lanzagorta, Stephen Bique, Robert Rosenberg
2009

On-Chip Networks
Natalie Enright Jerger, Li-Shiuan Peh
2009

The Memory System: You Can't Avoid It, You Can't Ignore It, You Can't Fake It
Bruce Jacob
2009

Fault Tolerant Computer Architecture
Daniel J. Sorin
2009

The Datacenter as a Computer: An Introduction to the Design of Warehouse-Scale Machines
Luiz André Barroso, Urs Hölzle
2009

Computer Architecture Techniques for Power-Efficiency
Stefanos Kaxiras, Margaret Martonosi
2008

Chip Multiprocessor Architecture: Techniques to Improve Throughput and Latency
Kunle Olukotun, Lance Hammond, James Laudon
2007

Transactional Memory
James R. Larus, Ravi Rajwar
2006

Quantum Computing for Computer Architects
Tzvetan S. Metodi, Frederic T. Chong
2006

© Springer Nature Switzerland AG 2022
Reprint of original edition © Morgan & Claypool 2014

All rights reserved. No part of this publication may be reproduced, stored in a retrieval system, or transmitted in any form or by any means—electronic, mechanical, photocopy, recording, or any other except for brief quotations in printed reviews, without the prior permission of the publisher.

A Primer on Hardware Prefetching
Babak Falsafi and Thomas F. Wenisch

ISBN: 9781-3-031-00615-9 print
ISBN: 9781-3-031-01743-8 ebook

DOI 10.1007/978-3-031-01743-8

A Publication in the Springer series
SYNTHESIS LECTURES ON ADVANCES IN AUTOMOTIVE TECHNOLOGY

Series Editor: Margaret Martonosi, Princeton University
Founding Editor Emeritus: Mark D. Hill, University of Wisconsin, Madison

Series ISSN 1935-3235 Print   1935-3243 Electronic

A Primer on Hardware Prefetching

Babak Falsafi
EPFL
Thomas F. Wenisch
University of Michigan

SYNTHESIS LECTURES ON COMPUTER ARCHITECTURE #28

ABSTRACT

Since the 1970's, microprocessor-based digital platforms have been riding Moore's law, allowing for doubling of density for the same area roughly every two years. However, whereas microprocessor fabrication has focused on increasing instruction execution rate, memory fabrication technologies have focused primarily on an increase in capacity with negligible increase in speed. This divergent trend in performance between the processors and memory has led to a phenomenon referred to as the "Memory Wall."

To overcome the memory wall, designers have resorted to a hierarchy of cache memory levels, which rely on the principal of memory access locality to reduce the observed memory access time and the performance gap between processors and memory. Unfortunately, important workload classes exhibit adverse memory access patterns that baffle the simple policies built into modern cache hierarchies to move instructions and data across cache levels. As such, processors often spend much time idling upon a demand fetch of memory blocks that miss in higher cache levels.

Prefetching—predicting future memory accesses and issuing requests for the corresponding memory blocks in advance of explicit accesses—is an effective approach to hide memory access latency. There have been a myriad of proposed prefetching techniques, and nearly every modern processor includes some hardware prefetching mechanisms targeting simple and regular memory access patterns. This primer offers an overview of the various classes of hardware prefetchers for instructions and data proposed in the research literature, and presents examples of techniques incorporated into modern microprocessors.

KEYWORDS

hardware prefetching, next-line prefetching, branch-directed prefetching, discontinuity prefetching, stride prefetching, address-correlated prefetching, Markov prefetcher, global history buffer, temporal memory streaming, spatial memory streaming, execution-based prefetching

Contents

Preface xiii

1 Introduction 1
   1.1 The Memory Wall 1
   1.2 Prefetching 3
       1.2.1 Predicting Addresses 3
       1.2.2 Prefetch Lookahead 4
       1.2.3 Placing Prefetched Values 4

2 Instruction Prefetching 7
   2.1 Next-Line Prefetching 7
   2.2 Fetch-Directed Prefetching 8
   2.3 Discontinuity Prefetching 10
   2.4 Prescient Fetch 12
   2.5 Temporal Instruction Fetch Streaming 12
   2.6 Return-Address Stack-Directed Instruction Prefetching 13
   2.7 Proactive Instruction Fetch 14

3 Data Prefetching 15
   3.1 Stride and Stream Prefetchers for Data 15
   3.2 Address-Correlating Prefetchers 17
       3.2.1 Jump Pointers 17
       3.2.2 Pair-Wise Correlation 18
       3.2.3 Markov Prefetcher 18
       3.2.4 Improving Lookahead via Prefetch Depth 19
       3.2.5 Improving Lookahead via Dead Block Prediction 20
       3.2.6 Addressing On-Chip Storage Limitations 21
       3.2.7 Global History Buffer 22
       3.2.8 Stream Chaining 24
       3.2.9 Temporal Memory Streaming 24
       3.2.10 Irregular Stream Buffer 25
   3.3 Spatially Correlated Prefetching 26
       3.3.1 Delta-Correlated Lookup 27

	3.3.2	Global History Buffer PC-Localized/Delta-Correlating (GHB PC/DC)	27
	3.3.3	Code-Correlated Lookup	28
	3.3.4	Spatial Footprint Prediction	30
	3.3.5	Spatial Pattern Prediction	30
	3.3.6	Stealth Prefetching	31
	3.3.7	Spatial Memory Streaming	31
	3.3.8	Spatio-Temporal Memory Streaming	32
3.4	Execution-Based Prefetching	33
	3.4.1	Algorithm Summarization	33
	3.4.2	Helper-Thread and Helper-Core Approaches	33
	3.4.3	Run-Ahead Execution	34
	3.4.4	Context Restoration	34
	3.4.5	Computation Spreading	35
3.5	Prefetch Modulation and Control	35
3.6	Software Approaches	36

4 Concluding Remarks **39**

Bibliography **41**

Author Biographies **53**

Preface

Since their inception in the 1970's, microprocessor-based digital platforms have been riding Moore's law, allowing for doubling of density for the same area roughly every two years. Microprocessors and memory fabrication technologies, however, have been exploiting this increase in density in two somewhat opposing ways. Whereas microprocessor fabrication has focused on increasing the rate at which machine instructions execute, memory fabrication technologies have focused primarily on an increase in capacity with negligible increase in speed. This divergent trend in performance between the processors and memory has led to a phenomenon referred to as the "Memory Wall" [1].

To overcome the memory wall, designers have resorted to a hierarchy of cache memory levels where at each level access latency is traded off for capacity. Caches rely on the principal of memory access locality to reduce the observed memory access time and the performance gap between processors and memory. Unfortunately, there are a number of important classes of workloads that exhibit adverse memory access patterns that baffle the simple policies built into modern cache hierarchies to move instructions and data across the cache levels. As such, processors often spend much time idling upon a demand fetch of memory blocks that miss in higher cache levels.

Prefetching—predicting future memory accesses and issuing requests for the corresponding memory blocks in advance of explicit accesses by a processor—is quite promising as an approach to hide memory access latency. There have been a myriad of hardware and software approaches to prefetching. A number of effective hardware prefetching mechanisms targeting simple and regular memory access patterns have been incorporated into modern microprocessors to prefetch instructions and data.

This primer offers an overview of the various classes of hardware prefetchers for instructions and data that have been proposed over the years, and presents examples of techniques incorporated into modern microprocessors. Although the techniques covered in this book are by no means comprehensive, they cover important instances of techniques from each class and as such the book serves as a suitable survey for those who plan to familiarize themselves with the domain. We cover prefetching for instruction and data caches, but many of the techniques we discuss may also be applicable to prefetching memory translations into translation lookaside buffers (see, e.g., [2]).

This primer is broken down into four chapters. In Chapter 1, we present an introduction to the memory hierarchy and general prefetching concepts. In Chapter 2, we describe techniques to prefetch instructions. Chapter 3 covers techniques to prefetch data, and we give concluding remarks in Chapter 4. The instruction prefetching techniques cover next-line prefetchers, branch-directed prefetching, discontinuity prefetchers, and temporal instruction streaming. The data prefetchers include stride and stream-based data prefetchers, address-correlated prefetching, spatially correlated prefetching, and execution-based prefetching.

We assume the reader is familiar with the basics of processor architecture and caches and has some familiarity with more advanced topics like out of order execution. This book enumerates the key issues in designing hardware prefetchers and provides high-level descriptions of a variety of prefetching techniques. We refer the reader to the cited publications for more complete microarchitectural details and performance evaluations of the prefetching schemes.

We would like to acknowledge those who helped contribute to this book. Thanks to Mark Hill for shepherding us through the writing and editorial process. Thanks to Cansu Kaynak and Michael Ferdman for providing figures and comments on drafts of this book. Thanks to Margaret Martonosi, Calvin Lin, and other anonymous reviewers for their detailed feedback that helped to improve this book.

CHAPTER 1

Introduction

1.1 THE MEMORY WALL

Figure 1.1 depicts the growing disparity between processor and memory performance in the past four decades. Innovations in microarchitecture, circuits, and fabrication technologies have led to an exponential increase in processor performance over this period. Meanwhile, DRAM has primarily benefitted from increases in density and DRAM speeds have improved only nominally. While future projections indicate that processor performance improvement may not continue at the same rate, the current gap in performance will necessitate techniques to mitigate long memory access latencies for years to come.

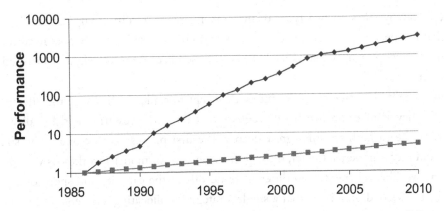

Figure 1.1: The growing disparity between processor and memory performance. From [3].

Computer architects have historically attempted to bridge this performance gap using a hierarchy of cache memories. Figure 1.2 depicts the anatomy of a modern computer's cache hierarchy. The hierarchy consists of cache memories that trade off capacity for lower latency at each level. The purpose of the hierarchy is to improve the apparent average memory access time by frequently handling a memory request at the cache, avoiding the comparatively long access latency of DRAM. The cache levels closer to the cores are smaller but faster. Each level provides a temporary repository for recently accessed memory blocks to reduce the effective memory access latency. The more frequently memory blocks are found in levels closer to the cores, the lower the access latency. We refer to the cache(s) closest to the core as the L1 caches and then number cache levels successively, referring to the final cache as the *last level cache* (LLC).

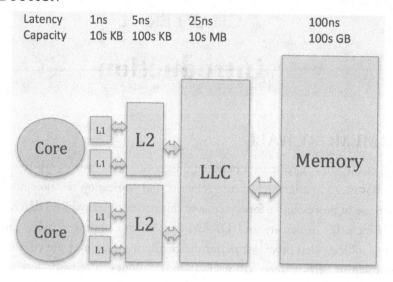

Figure 1.2: A modern memory hierarchy.

The hierarchy relies on two types of memory reference locality. Temporal locality refers to memory that has been recently accessed and is likely to be accessed again. Spatial locality refers to memory in physical proximity that is likely to be accessed because near-neighbor instructions and data are often related.

While locality is extremely powerful as a concept to exploit and reduce the effective memory access latency, it relies on two basic premises that do not necessarily hold for all workloads, particularly as the cache hierarchies grow deeper. The first premise is that one cache size fits all workloads and access patterns. In fact, the capacity demands of modern workloads vary drastically, and differing workloads benefit from different trade-offs in the capacity and speed of cache hierarchy levels. The second premise is that a single strategy for allocating and replacing cache entries (typically allocating on demand and replacing entries that have not been recently used) is suitable for all workloads. However, again, there is enormous variation in memory access patterns for which a simple strategy for deciding which blocks to cache may fare poorly.

There are a myriad of techniques that have been proposed from the algorithmic, compiler-level, and system software level all the way down to hardware to overcome the Memory Wall. These techniques include cache-oblivious algorithms, code and data layout optimizations at the compiler level, to hardware-centric approaches. Moreover, many software-based techniques have been proposed for prefetching. In this book, we focus on hardware-based techniques for prefetching instructions and data. For a more comprehensive treatment of the memory system, we refer the reader to the synthesis lecture by Jacob [4].

1.2   PREFETCHING

One way to hide memory access latency is to prefetch. Prefetching refers to the act of predicting a subsequent memory access and fetching the required values ahead of the memory access to hide any potential long latency. In the limit, a memory access does not incur any additional overhead and memory appears to have a performance equal to a processor register. In practice, however, prefetching may not always be timely or accurate. Late or inaccurate prefetches waste energy and, in the worst case, can hurt performance.

To hide latency effectively, a prefetching mechanism must: (1) predict the address of a memory access (i.e., be accurate), (2) predict when to issue a prefetch (i.e., be timely), and (3) choose where to place prefetched data (and, potentially, which other data to replace).

1.2.1   PREDICTING ADDRESSES

Predicting the correct memory addresses is a key challenge for prefetching mechanisms. If addresses are predicted correctly, the prefetching mechanism will have the opportunity to fetch them in advance and hide the memory access latency. If addresses are not predicted accurately, prefetching may cause pollution in the cache hierarchy (i.e., prefetched cache blocks would evict potentially useful cache blocks) and generate excessive traffic and contention in the memory system.

Predicting memory addresses may not be so simple. A data reference may be an access to a standalone variable or an element of a data structure and the nature of the reference depends on what the program is doing at a particular instance of execution. There are algorithms and data structure traversals that lend themselves well to both repetitive and predictable patterns (e.g., reading every element of an array sequentially). There are also a number of ways in which memory addresses can be hard to predict. These include, but are not limited to, interleaving of accesses to variables, multiple data structures, and control-flow dependent traversals (e.g., searching a binary tree).

Similarly, an instruction reference will depend on whether the program is executing sequentially or it is taking a branch (i.e., following a *discontinuity*). While sequential instruction fetch is straightforward, the control-flow behavior and its predictability in the program can impact how effective instruction prefetching can be.

Predicting addresses accurately also depends on the level of the cache hierarchy at which the prefetching is performed. At the highest level, the interface between the processor and level-one cache (Figure 1.2) contains all memory reference information that could enable highly accurate prefetch, but could also lead to a waste of resources recording prefetch information for accesses that will hit in the first level cache anyway, and thus do not require prefetch. Conversely, at lower hierarchy levels, the access sequence is filtered, observing only the misses from higher levels. Thus, otherwise effective prefetching algorithms may be confused by access-sequence perturbations from effects like cache placement and replacement policy.

Finally, there is typically a trade-off between the aggressiveness of a prefetch strategy and its accuracy; more aggressive prefetching will predict a higher fraction of the addresses actually requested by the processor at the cost of also fetching many more addresses erroneously. For this reason, many evaluation studies of prefetchers report two key metrics that jointly characterize the prefetcher's effectiveness at predicting addresses. *Coverage* measures the fraction of explicit processor requests for which a prefetch is successful (i.e., fraction of demand misses eliminated by prefetching). *Accuracy* measures the fraction of accesses issued by the prefetcher that turn out to be useful (i.e., fraction of correct prefetches over all prefetches). Many simple prefetchers can improve coverage at the expense of accuracy, whereas an ideal prefetcher provides both high accuracy and coverage.

1.2.2   PREFETCH LOOKAHEAD

Ideally, a prefetching mechanism issues a prefetch well in advance and provides enough storage for prefetched data so as to hide all memory access latency. Predicting precisely when to prefetch in practice, however, is a major challenge. Even if addresses are predicted correctly, a prefetcher that issues prefetches too early may not be able to hold all prefetched memory close to the processor long enough prior to access. In the best case, prefetching too early will be useless because the prefetched information will be evicted away from the processor prior to use. In the worst case, it may evict other useful information (e.g., other prefetched memory or useful blocks in higher-level caches). If memory is prefetched late, then it will diminish the effectiveness of prefetching by exposing the memory access latency upon the memory access. In the limit, late prefetches may lead to performance degradation due to additional memory system traffic and poor interaction with mechanisms designed to prioritize time-critical demand accesses.

1.2.3   PLACING PREFETCHED VALUES

The simplest and perhaps oldest software strategy for prefeching data is to load it into a processor register much like any other explicit load operation. Many architectures, in particular modern out-of-order processors, do not stall execution when a load is issued, but rather stall dependent instructions only when the value of a load is consumed by another instruction. Such a prefetch strategy is often called a *binding prefetch* because the value of subsequent uses of the data is bound at the time the prefetch is issued. This approach comes with a number of drawbacks: (1) it consumes precious processor registers, (2) it obligates the hardware to perform the prefetch, even if the memory system is heavily loaded, (3) it leads to semantic difficulties in the case the prefetch address is erroneous (e.g., should a prefetch of an invalid address result in a memory protection fault?), and (4) it is unclear how to apply this strategy to instructions.

Instead, most hardware prefetching techniques place prefetched values either directly into the cache hierarchy, or into supplemental buffers that augment the cache hierarchy, and are accessed concurrently. In multicore and multiprocessor systems, these caches and buffers participate in the cache coherence protocol, and hence the value of a prefetched memory location may change during the interval between the prefetch and a subsequent access; it is the hardware's responsibility to ensure the access sees the up-to-date value. Such prefetching strategies are referred to as non-binding. In these schemes, prefetching is purely a performance optimization and does not affect the semantics of a program.

All the hardware prefetchers we consider in this book fall into this latter, non-binding category. They differ in precisely where they place prefetched values and what they replaced to make room for newly prefetched memory.

CHAPTER 2

Instruction Prefetching

Instruction fetch stalls are detrimental to performance for workloads with large instruction working sets; when instruction supply slows down, the processor pipeline's execution resources (no matter how abundant) will be wasted. Whereas desktop and scientific workloads often exhibit small instruction working sets, conventional server workloads and emerging cloud workloads exhibit primary instruction working sets often far beyond what upper-level caches can accommodate. With trends towards fast software development, scripting paradigms, and virtualized environments with increasing software stack depth, primary instruction working sets are also growing fast. Modern hardware instruction scheduling techniques, such as out-of-order execution, are often effective in hiding some or all of the stalls due to data accesses and other long latency instructions. However, out-of-order execution generally cannot hide instruction fetch latency. As such, instruction stalls often account for a large fraction of overall memory stalls in servers.

2.1 NEXT-LINE PREFETCHING

Next-line prefetching [5] is the simplest form of instruction prefetching, which is prevalent in most modern processor designs. Because code is laid out sequentially in memory at consecutive memory addresses, often over half of lookups in the instruction cache are for sequential addresses. The logic needed to generate sequential addresses and fetch them is minimal and fairly easy to incorporate into a processor and cache hierarchy.

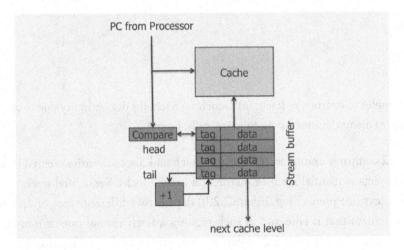

Figure 2.1: A next-line prefetcher.

Figure 2.1 depicts the anatomy of the next-line prefetcher in a modern processor pipeline. An instruction prefetch buffer or stream buffer, a small associative buffer, stores prefetched instruction cache blocks retrieved from lower cache hierarchy levels. Each time a block from the prefetch buffer is explicitly requested by the processor, it is transferred to the cache and the next consecutive block is prefetched from memory.

One of the first implementations of sequential instruction prefetching appeared in the IBM System/360 Model 91 in the late 1960's [6]. There are a number of research results pointing out the importance of next-line prefetching in server workloads [7, 8]. Many have also extended simple next-line prefetching schemes to arbitrary-length sequences of contiguous basic blocks [9, 10].

2.2   FETCH-DIRECTED PREFETCHING

Next-line prefetchers are quite effective and efficient but only half of instruction lookups are sequential. Control flow instructions break the sequential fetch and create discontinuities in fetch, and as such require predictability of future control flow and lookahead.

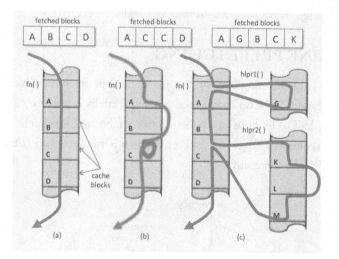

Figure 2.2: Examples of instruction fetch: (a) sequential fetch, (b) discontinuity due to an if-statement and a loop, and (c) discontinuities due to function calls. From [7].

Figure 2.2 compares examples of sequential fetch and discontinuities created by control flow. Figure 2.2(a) depicts sequential fetch of instruction cache blocks. Sequential fetch can be covered effectively with next-line prefetching. Figure 2.2(b) depicts two different types of discontinuity, one due to an if-statement that is false and as such requires a fetch around one or more cache blocks, and the other due to a loop. Figure 2.2(c) depicts discontinuities due to function calls.

Branch-predictor-directed prefetchers [11, 12, 13, 14, 15] reuse existing branch predictors to explore future control flow. These techniques use the branch predictor to recursively make future predictions to find instruction-block addresses for prefetch. Because branch predictors are, to the first order, decoupled from the rest of the pipeline, predictors can theoretically advance ahead of execution to an arbitrary extent to predict future control flow.

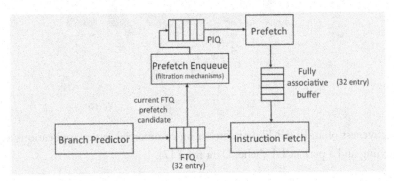

Figure 2.3: Fetch-directed instruction prefetching. From [12].

Fetch-directed instruction prefetching (FDIP) [12] is one of the best branch-predictor-directed techniques. Figure 2.3 shows the anatomy of FDIP. FDIP decouples the branch predictor from stalls in the L1 instruction fetch unit, introducing the *fetch target queue* (FTQ) between these two structures. The prefetcher uses the addresses in the FTQ to fetch instruction blocks from the L2 cache and place them in a small, fully associative buffer, overlapping prefetches with other L1 instruction fetches. The buffer is accessed by the instruction fetch unit in parallel with the L1 cache. To avoid redundancy between the buffer and L1, FDIP uses idle L1 instruction cache ports to probe the cache for the addresses in the FTQ to see if they are already present, and only enqueues missing addresses in the *prefetch instruction queue* (PIQ) for prefetch. Figure 2.4 illustrates the effectiveness of FDIP-like mechanisms for prefetching in commercial server applications, which incur substantial stalls due to instruction cache misses.

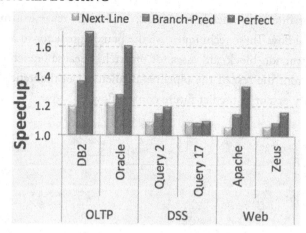

Figure 2.4: Effectiveness of an FDIP-like prefetcher on commercial server applications as compared to next-line prefetching and a perfect L1 cache. Data from [7].

Although effective at reducing instruction-fetch stalls, FDIP is fundamentally limited due to its limited prefetch look-ahead. Figure 2.5 quantifies the relationship between branch prediction bandwidth and prefetch lookahead. Nearly half of all instruction cache misses require in excess of 16 consecutive correct branch predictions (excluding inner-loop branches) before the candidate prefetch address can be generated.

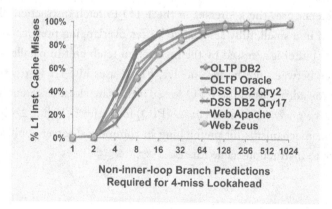

Figure 2.5: Correct branch predictions required to achieve 4-miss lookahead. Data from [7].

2.3   DISCONTINUITY PREFETCHING

A greater challenge lies in prefetching at fetch discontinuities—interruptions in the sequential instruction fetch sequence from function calls, taken branches, and traps.

There are a myriad of solutions to address control flow discontinuities. Wrong-path prefetching [16] is a simple approach to address the fundamental problem with FDIP by using the branch predictor, but predicting the opposite path. Although limited in its effectiveness, predicting the wrong path prefetches past data-dependent branches and through the exit of backward loop branches can prefetch instructions that FDIP cannot.

The branch-history guided prefetcher [17], execution-history guided prefetcher [18], multiple-stream predictor [10], next-trace predictors [19], and call graph prefetching [20] predict discontinuities keyed by earlier instructions, tracked independently from the branch predictor. Following the observation that server applications have deep call stacks of repeating functions, call graph prefetching [20] makes an effort to simultaneously predict the upcoming call stack rather just the next discontinuity.

One recent example of this approach is the discontinuity predictor [21], depicted in Figure 2.6, which maintains a table of fetch discontinuities mapping the PC of the block containing the taken branch to the branch target. Similar implementations are rumored to exist in some commercial processor products. As a next-line instruction prefetcher explores ahead of the fetch unit, it consults the discontinuity table with each block address and, upon a match, prefetches the discontinuous path in addition to the sequential one. Although it is simple and requires minimal hardware, the discontinuity predictor can bridge only a single fetch discontinuity; recursive lookups to explore additional paths result in an exponential growth in the number of prefetched blocks. The restriction to traverse at most one discontinuity limits the lookahead of the prefetcher. Furthermore, coverage is limited because the table will record only a single discontinuity per cache block, whereas there are cases where multiple taken branches occur within an instruction block.

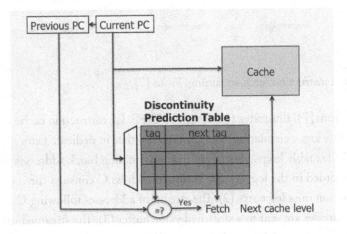

Figure 2.6: A discontinuity predictor.

2.4   PRESCIENT FETCH

Using idle or parallel resources has been proposed for instruction prefetching. Prescient fetch techniques [22, 23, 24, 25] use helper threads to identify critical computation and control transfers and perform them early, assisting the main thread that runs slower and in parallel with the helper thread. Although these approaches can be used for instruction prefetching beyond loops and function calls, only prescient instruction fetch [22] was specifically designed for this purpose. Speculative threading techniques identify the key execution information necessary and use it to forge ahead of the primary thread to issue instruction prefetches. Although speculative threading techniques can traverse multiple fetch discontinuities, their lookahead remains limited because they traverse the future instruction stream at the granularity of individual instructions, and hence often must traverse numerous instructions to discover a new cache block for prefetch.

2.5   TEMPORAL INSTRUCTION FETCH STREAMING

Temporal instruction fetch streaming (TIFS) [7] is designed to address the lookahead limitations of helper-thread and fetch/discontinuity-based mechanisms. Rather than explore a program's control flow graph, TIFS predicts future instruction-cache misses directly, through recording and replaying recurring L1 instruction miss sequences.

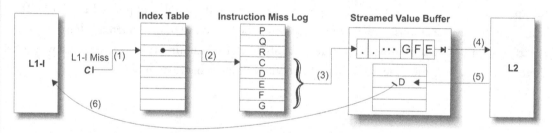

Figure 2.7: Temporal instruction fetch streaming. From [7].

Figure 2.7 (from [7]) illustrates the design of TIFS. L1 instruction cache misses are recorded in an instruction miss log, a circular buffer maintained either in dedicated storage or within the L2 cache. A separate index table keeps a mapping from instruction block addresses to the location that address was last recorded in the log. An L1-I miss to address C consults the index table (1), which points to an instruction miss log entry (2). The stream of addresses following C is read from the log and cache block addresses are sent to a streamed value buffer (3). The streamed value buffer requests the blocks in the stream from L2 (4), which returns the contents (5). Later, on a subsequent L1-I miss to D, the buffer returns the contents to the L1-I (6).

TIFS subsumes the sequential access predictions of next-line prefetchers. However, its performance benefit is greater because the predictions are more accurate and more timely. Increased accuracy comes from the fact that TIFS uses history to determine how many of the upcoming consecutive blocks should be prefetched.

TIFS improves lookahead in several ways. First, it operates at the granularity of cache blocks rather than individual instructions, addressing a key limitation of helper-thread approaches. Because of this, it skips over local loops and minor control flow within a cache block. TIFS is able to support any number of discontinuous branches and indirect branch targets by separately recording the discontinuities as part of the instruction stream. Furthermore, because it records extended sequences of instruction cache misses, it can quickly predict far into the future, providing substantially higher lookahead. For example, a next-line predictor is able to correctly prefetch a function body only after the first instruction block of that function is accessed. However, TIFS is able to predict and prefetch the same blocks earlier by predicting the function call and its sequential accesses prior to entering the function itself, while the caller is still executing code leading up to the call.

2.6 RETURN-ADDRESS STACK-DIRECTED INSTRUCTION PREFETCHING

Fetch-directed instruction prefetchers are limited due to the inability to predict past loop return branches and unpredictable conditional branches. Discontinuity prefetchers address these limitations, but rely on only a single PC value to predict an upcoming fetch discontinuity. Similarly, although TIFS increases lookahead, it still maintains only a single pointer from a cache block to a location in its log. As a result, both of these mechanisms lose accuracy when there are multiple control flow paths out of a particular cache block. Some common idioms, such as return instructions and switch statements, lead to multiple paths.

Return-address stack-directed instruction prefetching (RDIP) [26] uses additional program context information to enhance prediction accuracy and lookahead. RDIP is based on two observations: (1) program context as captured in the call stack correlates strongly with L1 instruction misses and (2) the *return address stack* (RAS), already present in all high performance processors, succinctly summarizes program context. RDIP associates prefetch operations with signatures formed from the contents of the RAS. It stores signatures and the associated prefetch addresses in a ~64 KB signature table, which is consulted upon each call and return operation to trigger prefetching. RDIP achieves 70% of the potential speedup of an ideal L1 cache, outperforming a prefetcherless baseline by 11.5% over a suite of server workloads.

2.7   PROACTIVE INSTRUCTION FETCH

In addition to its inability to distinguish program contexts, TIFS prediction accuracy is also hampered by other sources of control irregularity that cause slight differences in the sequence of L1 instruction cache misses. In particular, otherwise-repetitive instruction streams may be fragmented or filtered by small differences in cache replacements, the effects of instruction fetch along mispredicted branch paths, and the effects of asynchronous interrupts and operating system traps. *Proactive instruction fetch* [27] modifies the TIFS design to (1) record the sequence of cache blocks accessed by the committed instruction sequence (rather than instruction fetches that miss in the cache) and (2) separately record streams that execute in the context of interrupt/trap handlers. A key innovation of the design is a compressed representation of the instruction sequence that uses bit vectors to efficiently encode spatial locality among prefetch addresses. Follow-on work [28] centralizes the prefetcher metadata in a structure shared across cores, which substantially reduces storage cost in homogeneous designs where metadata is shared among many cores. Centralization reduces the per-core metadata to a minimal capacity, making the prefetcher practical even for many-core designs with small cores (e.g., cores used in mobile/embedded platforms).

Table 2.1: Summary of instruction prefetching techniques

Technique	Line of Attack	Lookahead	Accuracy	Cost/Complexity
Next-line	Sequential addresses	A few cache blocks	50%	< 1 KB
Fetch-directed	Follows the path of predicted control flow	Dependent on branch prediction accuracy	> 50%	< 1 KB
Discontinuity	Predicts discontinuities in control flow	Typically one branch ahead	> 50%	> 1 KB
Prescient Fetch	Helper threads	Limited by helper-thread execution bandwidth	> 50%	> 1 KB
Temporal streaming	Uses a single L1 miss to predict a sequence of L1 misses	An arbitrary number of cache blocks	95%	~64 K/core
RAS-directed	Context disambiguation for temporal streams	Same	> 95%	~64 K/core
Proactive Fetch	Uses L1 references to predict a stream of L1 misses	Same	> 99%	~256 K /chip

CHAPTER 3

Data Prefetching

Data miss patterns arise from the inherent structure that algorithms and high-level programming constructs impose to organize and traverse data in memory. Whereas instruction miss patterns in conventional von Neumann computer systems tend to be quite simple, following either sequential patterns or repetitive control transfers in a well-structured control flow graph, data access patterns can be far more diverse, particularly in pointer-linked data structures that enable multiple traversals. Moreover, whereas code tends to be static and hence easy to prefetch (with the exception of recent virtualization and just-in-time compilation mechanisms, which tend to thwart instruction prefetching), data structures morph over the course of execution, causing traversal patterns to change. This greater complexity in access patterns has led to a rich and diverse design space for data prefetching schemes that is much broader than instruction prefetchers.

We divide the design space of data prefetchers into four broad categories. First are prefetchers that rely on simple stride patterns, which directly generalize next-line instruction prefetching concepts to data. Second are those that rely on repetitive traversal sequences, often exploiting the pointer relationships among addresses. Third are those that rely on regular (yet potentially non-strided) data structure layouts. Finally, are mechanisms that explore ahead of the conventional out-of-order instruction window, and hence do not rely on regularity or repetition in the memory access address stream.

3.1 STRIDE AND STREAM PREFETCHERS FOR DATA

The first category of data prefetchers we examine are stride and stream prefetchers, which are a direct evolution of the next-line and stream prefetching mechanisms that have been developed for instructions. These prefetchers capture access patterns for data that are either laid out contiguously in the virtual address space or are separated by a constant stride. This class of prefetcher tends to be highly effective for dense matrix and array access patterns, but generally provides little benefit for pointer-based data structures. Strided data prefetchers are widely deployed in industrial processor designs, from systems as old as the IBM System/370 series through modern high-performance processors. Until recently, it is believed that this class of hardware data prefetcher was the only class to be commercially deployed.

Sequential data prefetcher implementations, which are restricted to prefetch only blocks at consecutive addresses, were described as early as 1978 [5]. By the early 1990's such prefetchers were extended to detect and prefetch sequences of accesses separated by a non-constant stride [29]. Such strided access patterns arise frequently when traversing multi-dimensional arrays or when aggregate data types (e.g., structs in C) are stored in arrays. Strided accesses can also arise by happenstance even in pointer-based data structures when dynamic memory allocators lay out constant-sized objects consecutively in memory, a common case due to pool allocators. Dahlgren and Stenstrom study the relative merits and effectiveness of sequential and stride prefetching mechanisms in detail [30].

A key challenge in stride prefetcher implementations is to distinguish among multiple interleaved strided sequences, for example, as may arise in a matrix-vector product. Figure 3.1 shows a diagram of Baer and Chen's scheme to track strides on a per-load-instruction basis. Their *reference prediction table* is a tagged, set-associative structure that uses the load instruction PC as the lookup key. Each entry holds the last address referenced by that load and the difference in address (i.e., stride) between the last two preceding references. Whenever the same stride is observed twice consecutively, the last reference address and stride are used to compute one or more additional addresses for prefetch. Subsequent accesses that continue to match the recorded stride will trigger additional prefetches. A long sequence of such strided accesses is referred to as a *stream*, analogous to instruction stream prefetchers. Ishii and co-authors describe more sophisticated hardware structures that can compactly represent multiple strides [31], while Sair and co-authors extend stream prefetching to more irregular patterns by predicting stride lengths [32].

Load Inst. PC (tag)	Last Address Referenced	Last Stride	Flags
....	....	......	

Figure 3.1: Baer and Chen's reference prediction table. From [29].

A second key implementation issue is to decide how many blocks to prefetch when a strided stream is detected. This parameter, often referred to as the *prefetch degree* or *prefetch depth*, is ideally large enough that the prefetched data arrive before being referenced by the processor, but not so large that blocks are replaced before access or cause undue pollution for short streams. Hur and Lin propose simple state machines that track histograms of recent stream lengths and can adaptively determine the appropriate prefetch depth for each distinct stream, enabling stream prefetchers to be effective even for short streams of only a few addresses [33].

Conventionally, stride prefetchers place the data they fetch directly into the cache hierarchy. However, if stride prefetchers are aggressive, they may pollute the cache, displacing useful data. Jouppi [34] describes an alternative organization wherein stream prefetchers place data in separate buffers, called *stream buffers*, which are accessed immediately after or in parallel with the L1 cache. By placing data in a stream buffer, a low accuracy stream (where many data are fetched but not used) does not displace useful data in the cache, reducing the risk of inaccurate prefetching. However, erroneous prefetches still consume energy and bandwidth. Palacharla and Kessler evaluate a memory system organization where stream buffers entirely replace the second-level data cache [35].

Each stream buffer holds cache blocks from a single stream. Accesses from the processor interrogate the stream buffer contents, typically in parallel with accesses to the L1 cache. A hit in a stream buffer typically causes the requested block to be transferred to the L1 cache and an additional block from the stream to be fetched. In some variants, stream buffers are strictly FIFO and only the head of each stream buffer may be accessed. In other variants, stream buffers are associatively searched. When the stride detection mechanism observes a new stream, an entire stream buffer is cleared and re-allocated (discarding any unreferenced blocks from a stale stream), typically according to a round-robin or least-recently-used scheme.

Additional implementation concerns and optimizations for stream prefetchers have been analyzed by Zhang and McKee [36] and Iacobovici and co-authors [37].

3.2   ADDRESS-CORRELATING PREFETCHERS

Whereas stride prefetchers are typically ineffective for pointer-based data structures, such as linked lists, the second class of prefetcher we consider is specifically designed to target the pointer-chasing access patterns of such data structures. Instead of relying on regularity in the layout of data in memory, this class of prefetcher exploits the fact that algorithms tend to traverse data structures in the same way repeatedly, leading to recurring cache miss sequences.

Correlation between accesses to pairs of memory locations was suggested as early as 1976 [38]. Charney and Reeves first described hardware prefetchers that seek to exploit such pair-wise correlation relationships, coining the term "correlation-based prefetcher" [39, 40]. Later work generalizes the notion of address correlation from pairs to groups or sequences of accesses [41, 42]. Wenisch and co-authors introduce the term "temporal correlation" [42] to refer to the phenomenon that two addresses accessed near one another in time will tend to be accessed together again in the future. Temporal correlation is an analog to "temporal locality," that a recently accessed address is likely to be accessed again in the near future. Whereas caches exploit temporal locality, address-correlating prefetchers exploit temporal correlation.

3.2.1   JUMP POINTERS

Correlating prefetchers are a generalization of hardware and software mechanisms that specifically targeted pointer-chasing access patterns. These earlier mechanisms rely on the concept of a *jump pointer* [43, 44, 45, 46], a pointer that enables a large forward jump in a data structure traversal. For example, a node in a linked list may be augmented with a pointer ten nodes forward in the list; the prefetcher can follow the jump pointer to gain lookahead over the main traversal being carried out by the CPU, enabling timely prefetch. Prefetchers relying on jump pointers often require software or compiler support to annotate pointers. *Content directed prefetchers* [47, 48] eschew annotation and attempt instead to dereference and prefetch any load value that appears to form a valid virtual address. While jump-pointer mechanisms can be quite effective for specific data structure traversals (e.g., linked list traversals), their key shortcoming is that the distance the jump pointer advances the traversal must be carefully balanced to provide sufficient lookahead without jumping over too many elements. Jump pointer distances are difficult to tune and the pointers themselves can be expensive to store.

3.2.2 PAIR-WISE CORRELATION

In essence, a correlation-based hardware prefetcher is a lookup table that maps from one address to another address that is likely to follow it in the access sequence. While such an association can capture sequential and stride relationships, it is far more general, capturing, for example, the relationship between the address of a pointer and the address to which it points. It is the ability to capture pointer traversals that affords address-correlating prefetchers a far greater opportunity for performance improvement than stride prefetchers, as pointer-chasing access patterns are disproportionately slow on modern processors. However, address-correlating prefetchers rely on repetition; they are unable to prefetch addresses that have never previously been referenced (in contrast to stride prefetchers). Moreover, address correlation prefetchers require enormous state, as they need to store the successor for every address. Hence, their storage requirement grows proportionally to the working set of the application. Much of the innovation in address-correlating prefetcher design centers on managing this enormous state.

3.2.3 MARKOV PREFETCHER

The *Markov prefetcher* [49, 50] is the simplest prefetcher design to exploit pair-wise address correlation. It directly implements the notion of a look-up table mapping a trigger address to its immediate successor in the off-chip access sequence. However, because addresses—especially when considered at cache-block granularity—often participate in multiple traversals, storing only a single successor for each trigger address results in poor effectiveness. Instead, the Markov prefetcher stores several previously observed successors, all of which are prefetched when a miss to the trigger address is observed. By prefetching several possible successors, the Markov prefetcher sacrifices accuracy (fraction of correct prefetches over all prefetches) to improve coverage (fraction of demand misses for which a prefetch is successful)—only one of the addresses requested by the prefetcher is expected to be correct. The number of successors fetched is often referred to as the *width* of the prefetch.

The Markov prefetcher is organized as an on-chip set-associative table indexed by trigger address (see Figure 3.2). An entry in the table contains the set of successor addresses to prefetch and optional confidence or replacement policy information; four successors is a typical width. The original study [49] proposed 1 MB lookup tables. However, the required lookup table size grows with applications' data footprints, and follow-on studies have shown that modern workloads require far larger tables for effective prefetching [51].

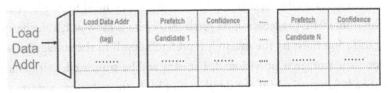

Figure 3.2: Markov prefetcher, described in [49].

The Markov prefetcher design is inspired by conceptualizing a Markov model of the off-chip access sequence. Each state in the model corresponds to a trigger address, with possible successor states corresponding to subsequent miss addresses. Transition probabilities in the first-order Markov model correspond to the likelihood of each successor miss. The objective of the lookup table is to store the successors with the highest transition probabilities for the most frequently encountered triggers. However, existing hardware proposals do not explicitly calculate trigger or transition probabilities; both the trigger addresses and the successors for each are managed heuristically using least-recently used (LRU) replacement.

Two factors limit the effectiveness of Markov prefetchers: (1) lookahead and memory-level-parallelism are limited because the prefetcher attempts to predict only the next miss and (2) coverage is limited by on-chip correlation table capacity. We next discuss several proposals to address each of these limitations.

3.2.4 IMPROVING LOOKAHEAD VIA PREFETCH DEPTH

The first key limitation of Markov prefetchers are their limited lookahead—that is, the time between the trigger address and the stall on the subsequent address is too short to hide the access latency of the prefetched data.

A straightforward approach to improve lookahead is to fetch more addresses, further ahead in the predicted global address sequence [41], analogous to increasing the depth of a stream prefetcher. With a Markov-like prefetcher organization, deeper prefetches can be issued by recursively performing table lookups using the initial set of predicted addresses. However, such an approach incurs high lookup latency and drastically increases bandwidth demands on the Markov table. Moreover, the number of possible prefetch candidates grows geometrically with prefetch depth, so an appropriate policy for limiting this growth is needed to maintain accuracy.

Alternatively, rather than perform consecutive lookups, the prefetch table can be folded to store a short sequence (a.k.a. stream) of successors alongside each prefetch trigger [52, 53]. An appropriate policy is still needed to select which successor stream(s) to record, for example, the most recent or highest probability successors. This approach solves the lookup latency/bandwidth problem, but can complicate table update, since an address may need to be recorded in several entries (e.g., for the immediate predecessor, its predecessor, etc.). However, a deeper problem is that such a table organization must fix the maximum prefetch depth to the storage provisioned in each table entry. If too little storage is provisioned, a stream will be truncated, sacrificing potential coverage and lookahead. Alternatively, if the depth is too high, then storage is wasted or, even worse, accuracy may suffer as uncorrelated addresses are recorded at the tail of the stream. Several studies have shown that the length of repetitive streams varies over orders of magnitude, from as few as two to many thousands of misses [41, 42, 52, 54, 55]. We discuss alternative storage organizations to address this problem later.

Typically, the first few misses at the start of a stream are not timely—the trigger miss is simply too close in time to the accesses of the subsequent blocks to allow timely prefetch. Recording and prefetching these addresses wastes storage and delays prefetch of the useful blocks deeper in a stream. Hence, it makes sense to omit these first few addresses, and begin the stream with the first miss for which prefetching can provide a latency advantage. Chou and co-authors observe that, in out-of-order cores, the instruction window frequently causes several off-chip misses to issue concurrently [56]. Execution then stalls while this group of misses is serviced. Once these misses return, the instruction window can advance and dependent addresses can be calculated, allowing the next group of misses to issue in parallel. They refer to the average number of misses issued in each of these groups as the *memory level parallelism* of an application. Chou proposes an *epoch-based correlation prefetching* (EBCP) mechanism that exploits this observation [57]. In EBCP, the first miss within each parallel group (epoch) is used as a trigger address and is used to look up misses in the next (or subsequent) groups, thereby skipping over miss addresses that are likely already in flight when the trigger miss is encountered.

3.2.5   IMPROVING LOOKAHEAD VIA DEAD BLOCK PREDICTION

A second approach to address the lookahead limitation of the Markov prefetcher is to select an earlier trigger event for each prefetch operation. *Dead block prediction* [51, 54, 60, 61, 62] is based on the key observation that cache blocks spend a majority of their time in the cache *dead* [63, 64, 65]—that is, they remain cached but will not be accessed again prior to invalidation or eviction. Dead cache blocks occupy storage space but will provide no further cache hits. As such, they present an opportunity: a prefetcher can replace the dead block with a prefetched block with no risk of cache pollution. A *dead block correlating prefetcher* (DBCP) [51] seeks to predict the last access to a cache block (i.e., its death event) and then use this access as a trigger to prefetch the block that will replace the dead block. In one sense, DBCP maximizes lookahead, as it issues a prefetch at the earliest moment that storage in the cache becomes available to receive the prefetched block.

DBCP relies on two predictions. First, it must predict when a cache block becomes dead. Death events can be predicted based on *code correlation* [51, 54, 62], or *time keeping* [61]. Code-correlated dead block prediction seeks to recognize the last instruction to access a cache block prior to its eviction, i.e., the access upon which the block becomes dead. Code correlation relies on the observation that cache blocks tend to be accessed by the same sequence of loads and stores each time they enter the cache, from the initial miss that allocates the block, through a sequence of accesses, and ultimately to the last access when the block becomes dead. A key advantage of code correlation is that an access sequence learned for one cache block can be applied to predict death events for other addresses. We discuss code correlation in detail Section 3.3.3.

Alternatively, time-keeping mechanisms seek to predict the time until a cache block dies rather than the specific access indicating its death [61]. The observation underling this approach is that the lifetime of a block (measured in clock cycles) tends to be similar each time it enters the cache. In such designs, the Markov prefetch table is augmented with an additional field indicating the lifetime of the block the last time it entered the cache. The block is then predicted as dead after some suitable safety margin (e.g., double the prior lifetime).

Once a death event has been predicted, a suitable prefetch candidate to replace it must be predicted. Early dead block prefetchers used Markov-like prediction tables for this purpose [51, 61]. Later proposals [54] rely on more sophisticated temporal stream predictors, which we describe in Section 3.2.9.

3.2.6   ADDRESSING ON-CHIP STORAGE LIMITATIONS

The second key aspect of Markov prefetchers that limits their effectiveness is the limited on-chip storage capacity of the correlation table. Fundamentally, the size of the meta-data required for address correlation grows with the working set of an application; ideally the correlation table captures correlations for all addresses within the working set. The required correlation table size is thus a constant factor smaller than the data set itself (i.e., since it stores addresses rather than data, the correlation table storage requirement is smaller than the working set by the ratio of cache block size to address size).

One approach to improve Markov prefetcher effectiveness is to improve the storage efficiency of the on-chip correlation table. The *tag correlating prefetcher* [66] stores only the tag rather than the complete cache block address in correlation table entries. This modification reduces the storage requirement per entry, but still improves correlation table capacity by only a small constant factor, which is insufficient to obtain high coverage for applications with large working sets (e.g., server applications).

A second approach is to relocate the correlation table to main memory, eliminating the capacity restrictions of an on-chip correlation table [42, 52, 54, 67, 72]. Off-chip correlation tables can achieve high coverage, even for workloads with large working sets [52]. However, shifting the correlation table off chip increases the access latency for prefetcher meta-data from a few clock cycles to the latency of an off-chip memory reference—accessing the prefetch meta-data thus may take as long as prefetching the desired cache block.

To be effective, off-chip correlation tables must provide sufficient lookahead to hide the long meta-data access latency. One approach, discussed in Section 3.2.4, is to design the prefetcher to record addresses for memory references that will occur in a future parallel group (epoch), as in EBCP [57]. A second approach is to increase prefetch depth [52]; even if the first block to be prefetched is not timely, later addresses will be successfully prefetched. A generalization of increasing prefetch depth is *temporal streaming* [42], discussed in Section 3.2.9, which amortizes off-chip meta-data references by targeting streams of arbitrary length (i.e., effectively unlimited prefetch depth) [67].

3.2.7   GLOBAL HISTORY BUFFER

The cache-like organization of a Markov prefetcher limits it to record only fixed length streams—a single table entry can store addresses for only a fixed prefetch depth. Narrow table entries sacrifice potential coverage and lookahead, while wide table entries are storage-inefficient for short streams. Entries can be chained together via pointers, however this increases lookup latency and is particularly undesirable if correlation tables are located off chip. Wenisch and co-authors study repetitive temporally correlated streams in commercial server applications and demonstrate that stream lengths vary from two to many thousands of cache blocks [55, 67]. The most common stream length is only two misses, implying that a wide Markov table entry is storage inefficient. However, when weighted by the number of misses in the stream (i.e., the potential coverage that can be obtained by prefetching the stream), the median stream length is about ten cache blocks.

A key advance, introduced by Nesbit and Smith in their *global history buffer* [68], is to split the correlation table into two structures: a *history buffer*, which logs the sequence of misses in a circular buffer in the order they occurred, and an *index table*, which provides a mapping from an address (or other prefetch trigger) to a location in the history buffer. The history buffer allows a single prefetch trigger to point to a stream of arbitrary length. Figure 3.3 (from [68]) illustrates the Global History Buffer organization.

The index table retains a set-associative storage organization similar to the original Markov prefetcher. However, rather than storing cache block addresses, the index table now stores pointers into the history buffer. When a miss occurs, the GHB references the index table to see if any information is associated with the miss address. If an entry is found, the pointer is followed and the history buffer entry is checked to see if it still contains the miss address (the entry may since have been overwritten). If so, the next few entries in the history buffer contain the predicted stream. History buffer entries can also be augmented with link pointers to other history buffer locations, to enable history traversal according to more than one ordering (e.g., each link pointer may indicate a preceding occurrence of the same miss address, enabling increases to prefetch width as well as depth).

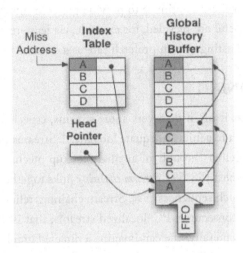

Figure 3.3: Address-correlating global history buffer (GHB G/AC). From [68].

By varying the key stored in the index table and the link pointers between history buffer entries, the GHB design can exploit a variety of properties that relate trigger events to predicted prefetch streams. Nesbit and Smith introduce a taxonomy of GHB variants of the form GHB *X/Y*, where X indicates how streams are localized (i.e., how link pointers connect history buffer entries that should be prefetched consecutively) and Y indicates the correlation method (i.e., how the lookup process locates a candidate stream) [68, 69]. Localization can be *global* (G) or *per-PC* (PC). Under global localization, consecutively recorded history buffer entries form a stream. The pointer associated with each history table entry either points to earlier occurrences of the same miss address (facilitating higher prefetch width as discussed above) or is unused. Under per-PC localization, both the index table and link pointers connect history buffer entries based on the PC of the trigger access; a stream is formed by following the link pointers connecting consecutive misses issued by the same trigger PC. The correlation method may be *address correlating* (AC) or *delta correlating* (DC). In this section, we discuss the global address correlating variant (GHB G/AC), where the index table maps miss addresses to history buffer locations. In Section 3.3.2, we discuss GHB PC/ DC ("program counter-localized delta correlation"), which instead locates entries and records history based on the stride between consecutive misses and localizes miss histories on a per-PC basis. The literature discusses several other alternatives for localization and correlation [68, 69].

One challenge under the GHB organization is to determine when a stream ends, that is, when the prefetcher should no longer fetch additional addresses indicated in the history buffer. Many proposals that build on the GHB organization (e.g., [42]) make no effort to predict the end of a stream. Instead, they allocate a stream buffer [34] for each successful index table lookup and continue follow the stream while it continues to provide prefetch hits. Stream buffers allocated to streams that are no longer useful are recycled, for example, via least-recent-used replacement. Wenisch discusses adaptively adjusting stream prefetch rate as a stream is followed [42].

3.2.8   STREAM CHAINING

A limitation of GHB is that it often discovers short streams, especially when localizing streams per-PC, for which it is difficult to achieve adequate lookahead. Streams can be extended by chaining separately stored streams together through an auxiliary lookup mechanism that predicts relationships from one stream to the next [58, 59]. *Stream chaining* links together streams formed by different PCs to create longer prefetch sequences [59]. Stream chaining relies on the insight that there is temporal correlation among consecutive PC-localized streams; that is, the same two streams tend to recur consecutively. More generally, one can imagine a directed graph among streams indicating which subsequent stream is most likely to follow each predecessor stream. Stream chaining extends each index table entry with a *next stream* pointer that indicates the index table entry that was used next after this stream, along with a confidence counter. By linking together individual PC-localized streams, stream chaining enables more correct prefetches from a single trigger access.

3.2.9   TEMPORAL MEMORY STREAMING

As originally proposed, the GHB index and history tables are small and located on chip, thus limiting the effectiveness of the GHB/AC variant for workloads with large working sets. Temporal Memory Streaming [42] adopts the GHB storage organization, but places both the index table and history buffer off chip in main memory, allowing it to record and replay arbitrary length streams even for workloads with large working sets.

Off-chip tables introduce two challenges. We have discussed the meta-data access latency and prefetch lookahead challenges in Section 3.2.6. However, updating off-chip tables also presents a challenge. Since the tables must be updated on every miss, a naïve implementation would triple memory bandwidth (one memory access for the miss, and one each for the index table and history buffer updates).

History table update bandwidth is easily addressed by maintaining a small buffer on chip and coalescing consecutive history table appends into a single write. Index table updates do not benefit from spatial locality and cannot exploit the same optimization. Wenisch and co-authors show, however, that index table update bandwidth can be managed by sampling the index table updates, performing only a random subset of history table writes [67, 70]. Their study shows that streams that account for the majority of coverage are either long or recur frequently. For long streams, an index table entry is recorded with high probability within a few accesses of the start of the stream, sacrificing negligible coverage relative to the long body of the stream. For frequent streams, even though an index table entry may not be recorded the first time the stream is traversed, the probability of recording the desired entry rapidly approaches one as the stream recurs. Figure 3.4 (from [67]) illustrates why sampling is effective.

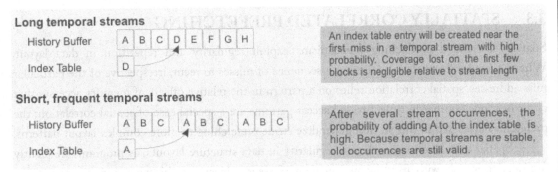

Figure 3.4: Sampling index table updates is effective for long and short, frequent streams. From [67].

IBM recently announced that the IBM Blue Gene/Q includes a new prefetching scheme, called *list prefetching* [71], that bears many similarities to temporal memory streaming and is, to our knowledge, the only publicly disclosed commercial implementation of such a prefetcher. The list prefetching engine can prefetch from a recorded miss stream located in main memory. The address list can either be provided via a software API or recorded automatically by hardware. However, the list prefetcher does not provide an off-chip index table; software must assist the prefetcher in recording, locating and initiating streams. Hardware then manages timeliness of prefetch and small deviations between the recorded stream and L1 misses.

3.2.10 IRREGULAR STREAM BUFFER

Whereas sampling can reduce the overheads of maintaining stream meta-data off-chip, lookup latency remains high, limiting prefetch lookahead for short streams. Furthermore, off-chip storage precludes GHB-like PC-localization of address-correlated streams; following link pointers between entries in an off-chip history buffer is too slow (i.e., it is no faster than dereferencing a chain of dependent pointers, precisely the access pattern temporal streaming is designed to accelerate). To address these limitations, Jain and Lin introduce the *irregular stream buffer* (ISB) [72]. The ISB introduces a new conceptual address space, the *structural address space*, which is visible only to the prefetcher. As cache blocks are accessed in the last-level cache, they are assigned consecutive structural addresses (potentially replacing a previous structural address assignment for a particular cache block). With this remapping, temporal correlation in the physical address space becomes spatial correlation in the structural address space. Thus, a temporal stream of physical addresses can be prefetched with a simple next-n-line stream prefetcher that issues requests to structural addresses. Two on-chip tables maintain a bidirectional mapping between physical and structural addresses. These mappings are shifted from the on-chip tables to an off-chip backing store in tandem with fills and replacements in the TLB, ensuring that the on-chip structures contain meta-data for the set of addresses the processor can presently access.

3.3   SPATIALLY CORRELATED PREFETCHING

Spatially correlated prefetching mechanisms exploit regularity and repetition in data layout. Whereas temporal correlation relies on a sequence of misses to recur, irrespective of the particular miss addresses, spatial correlation relies on a pattern in the relative offsets of memory accesses that occur near one another in time. Strided access patterns are a special case of spatial correlation; the prefetchers discussed in this section generalize stride prefetching to more complex layout patterns.

Spatial correlation arises due to regularity in data structure layouts, as programs frequently use data structures with fixed layouts in memory (e.g., objects in object-oriented programming language, database records with fixed-size fields, stack frames). Many high-level programming languages align data structures to cache-line and page boundaries, further increasing layout regularity.

One of the strengths of spatial correlation is that the same layout patterns often recur for many objects in memory. Hence, once learned, a spatial correlation pattern can be used to prefetch many objects, making prefetchers that exploit spatial correlation highly storage efficient (i.e., a compact pattern can be reused frequently to prefetch many addresses). Furthermore, in contrast to address correlation, which relies on repetition of misses to particular addresses, spatial patterns can be applied to addresses that have never previously been referenced, and hence can eliminate cold cache misses.

Like address correlating prefetchers, spatial correlating prefetchers must rely on a trigger event to initiate prefetch, which is usually a memory access or cache miss. The trigger event must (1) provide a key to lookup the relevant layout pattern describing the relative locations to prefetch and (2) provide the base address from which to calculate those relative addresses. The base address is usually obtained from the effective address of the triggering access. However, to be able to prefetch for previously unseen addresses, the lookup key for the layout pattern must be independent of the base address (if the lookup key is simply the base address, for example, as in GHB G/AC [68], then we would classify the prefetcher as address correlating).

3.3.1   DELTA-CORRELATED LOOKUP

Delta correlation builds directly on the notion of spatial correlation to exploit the repetition in the layout pattern itself as the lookup key—delta correlation uses the prefix of the layout pattern as the lookup key. That is, the stride between two or more consecutive accesses (or a signature summarizing a sequence of such strides) is used as the lookup key for the layout pattern. Delta correlation (referred to as distance prefetching in early works) was originally proposed in the context of TLB prefetching [73].

3.3.2   GLOBAL HISTORY BUFFER PC-LOCALIZED/DELTA-CORRELATING (GHB PC/DC)

One of the most effective known prefetchers from both a storage efficiency and coverage perspective is the program counter-localized delta-correlating variant of the global history buffer (GHB PC/DC) [68, 69]. The hardware organization of GHB PC/DC is the same as that of GHB G/AC (global address-correlating variant of GHB) discussed in Section 3.2.7, comprising an index table and a history table. The key stored in the GHB PC/DC index table is the program counter value of the missing memory access instruction. Consecutive misses by the same PC are linked together via link pointers stored in the history buffer. Upon a miss by a given PC, the history table is accessed to identify the preceding two misses from the same PC, navigating from one miss to the next via the link pointers. The deltas (strides) between the trigger and preceding two misses are calculated by subtracting the miss addresses, producing a two-stride sequence. The prefetcher then continues searching backwards in the history buffer following the PC link pointers to search for a preceding occurrence of the same two strides. If such a pattern is found, prefetch addresses are calculated by applying the subsequent delta sequence, starting at the match, to the base address of the trigger miss. This entire search process is implemented with a state machine that traverses the index and history tables.

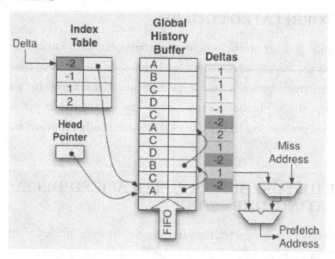

Figure 3.5: Delta-correlating global history buffer (GHB G/DC). From [68].

Although GHB PC/DC has demonstrated remarkable effectiveness with only limited storage (256-entry index and history buffer tables) for SPEC benchmarks, to date, its effectiveness has not been studied with workloads that have large code and data footprints (such as commercial server or cloud computing applications) and its scaling behavior is unknown.

Several innovative variants of the GHB PC/DC prefetcher were evaluated in the First Data Prefetching Championship and described in a special issue of the *Journal of Instruction Level Parallelism* in 2011 [31, 74, 75, 76, 77, 78].

3.3.3   CODE-CORRELATED LOOKUP

PC-localized stride/stream predictors and the high effectiveness of the PC-localized GHB demonstrate that the spatial relationships between particular memory accesses are often strongly correlated to the memory access instruction PC. For example, a particular load in the inner loop of a matrix multiply will frequently traverse memory with a constant stride. Code correlation generalizes the notion of PC-localized streams by observing that complex traversal patterns, often comprising several memory access instructions, are also strongly correlated to the PCs of the issuing instructions. That is, a particular instruction sequence accesses memory with the same cookie-cutter-like offset pattern repeatedly at many different base addresses in memory. Such patterns arise because stack frames and objects in object-oriented programs have fixed layouts; executing the code that establishes the stack frame or invoking a particular method on the object is a reliable indication of the upcoming pattern of memory accesses to fields within the stack frame/object.

Code-correlated prediction was first explored in the context of mechanisms that predict cache coherence state machine transitions [60, 79, 80]. We previously introduced code correlation in the context of dead-block prediction for improving address correlating prefetch lookahead in Section 3.2.5.

Code-correlated spatial prefetchers associate spatial relationships with the program counter value of the trigger event (rather than the accessed data addresses or deltas among them). The general architecture for such prefetching mechanisms is shown in Figure 3.6 (from [82]). The trigger event for prefetching is the first access or miss to a particular (usually fixed-size) region of memory. Region sizes vary across specific prefetcher designs and range from as small as 256 bytes [81] to as large as 8 KB [82].

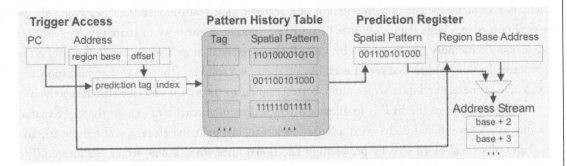

Figure 3.6: Code-correlated spatial prefetching. From [82].

Upon a trigger event (e.g., an L1 cache miss), the prefetcher constructs a lookup key from the trigger event and searches for this key in a *pattern history table*, which associates the key with a *spatial pattern*, a representation of the relative offsets to prefetch. Trigger events, lookup keys, spatial pattern encoding, pattern history table organization, and the mechanisms used to train the prefetcher vary among specific prefetcher designs.

The lookup key typically includes some or all of the bits from the PC of the trigger access. Several studies [81, 82, 83] show that additionally including low-order bits of the data address, in particular, the offset within the region, improves prefetch accuracy. These low-order data address bits serve to distinguish among accesses to objects with similar layouts that are aligned differently with respect to region boundaries—separate entries are recorded in the predictor tables for each possible alignment. Alternatively, Ferdman and co-authors propose storing only a single pattern using the PC as the lookup key and instead use the low-order bits to rotate the pattern to the appropriate alignment [84, 85].

The simplest representation for spatial patterns is a bit vector representing which portions (e.g., cache lines) of the region should be prefetched. This bit vector, combined with the base address of the region (taken from the data address requested by the trigger event), provides the list of addresses to prefetch.

In the following subsections, we briefly summarize key aspects of specific code-correlated prefetcher designs.

3.3.4   SPATIAL FOOTPRINT PREDICTION

The earliest and simplest code-correlated spatial prefetcher is the *spatial footprint predictor* (SFP) of Kumar and Wilkerson [81]. They explore repetitive layouts of L1 data cache accesses in the context of a decoupled sectored cache [86], a tag organization that facilitates small line size with low overhead by breaking the one-to-one association of tags and data. The decoupled sectored cache is a storage-efficient implementation of a sectored (a.k.a sub-blocked) cache, which allows several entries in a large data array that lie near one another in memory to share tags in a smaller tag array. The memory block covered by a single tag is referred to as a sector, while the data array stores smaller lines within that sector each having their own valid bit. Thus, in this organization, the block size of the tag and data arrays differ, corresponding to the sector and line size, respectively.

The objective of the SFP is to allow a decoupled sectored cache to exploit the same spatial locality as a conventional cache with a large block size, but gain the storage and bandwidth advantages of a small block size, by prefetching additional lines when a new sector is allocated. The authors study a system where the 16KB L1 data cache is a decoupled sectored cache with 128-byte sectors but 8-byte line size with storage for 2048 data array entries but only 512 tags.

As in the generic architecture shown in Figure 3.6, the best-performing SFP variant stores its predictions in a set-associative pattern history table indexed and tagged with the PC and line number within the sector. The pattern history table is trained by extending each sector's tag entry with additional storage for the PC of the miss that caused the sector to be allocated. When a sector is replaced, this PC and the valid bit vector of lines within the sector are stored in the pattern history table.

3.3.5   SPATIAL PATTERN PREDICTION

Chen and co-authors describe the *spatial pattern predictor* (SPP) [83] and show that it can be used to prefetch 64 B cache blocks within larger regions up to 1 KB in size in the context of a conventional set-associative cache. A surprising result of this study is that spatial pattern prediction can be tuned to require remarkably little storage; a 256-entry tag-less predictor (well under 1 KB total storage) can provide 95% coverage with less than 20% erroneous prefetches for SPEC applications. The study further shows how spatial pattern prediction can be combined with circuit techniques [87] to reduce cache leakage energy for portions of a cache line that are unlikely to be accessed.

3.3.6   STEALTH PREFETCHING

Cantin and co-authors propose a spatial prefetching technique, *stealth prefetching* [88], that is specifically targeted for broadcast snooping bus-based multiprocessors. The objective of stealth prefetching is to design a spatial prefetcher that avoids the negative cache coherence implications of aggressively prefetching shared memory blocks. The technique relies on coarse-grain spatial tracking of shared and private memory regions and prefetches data from private regions into dedicated storage while skipping many of the costly operations involved in cache coherent loads; the coherency of the prefetched blocks follows from the safety guarantees of the underlying region tracking tracking technique. The key advantage of stealth prefetching is that the prefetch operations do not cause disruptive snooping operations at other caches.

3.3.7   SPATIAL MEMORY STREAMING

Whereas SFP and SPP target only the L1 miss stream, fetching most data from L2, Somogyi and co-authors demonstrate that code-correlated spatial prefetching is also effective for off-chip misses [82]. Their study is also the first to demonstrate the effectiveness of spatially correlated prefetching in commercial server applications.

The SMS design employs 8 KB region sizes, and includes critical optimizations to facilitate storage-efficient training of the pattern history table. This additional structure is required because SMS tracks spatial patterns for a much larger memory footprint (all on chip caches) than SFP or SPP (L1 cache only). Furthermore, whereas SPP and SFP track only a single spatial pattern for a group of adjacent cache frames, the high associativity of the L2 cache makes it important for SMS to be able to track multiple spatial regions that are concurrently resident in the cache. To this end, SMS introduces the notion of a *spatial region generation* that begins with the first miss to access a block within a region and ends with the eviction or invalidation of any block from that region. SMS's training mechanism tracks spatial patterns over an entire generation. Patterns are accumulated in the *active generation table*, shown in Figure 3.7. A critical optimization of the active generation table is the use of the *filter table* to reduce storage requirements. When a miss initiates a new spatial region generation, the corresponding tag and trigger PC/offset are first placed in the filter table, which is managed with LRU replacement. Only upon a second miss to the region are filter table entries transferred to the *accumulation table*, which tracks the spatial region generation until eviction of any block, at which point the spatial pattern is recorded in the pattern history table common to all code-correlated prefetcher designs. The use of the filter table is important because the majority of spatial region generations are singleton misses, containing only the trigger miss; the filter table ameliorates the storage pressure these useless entries would otherwise cause in the accumulation table.

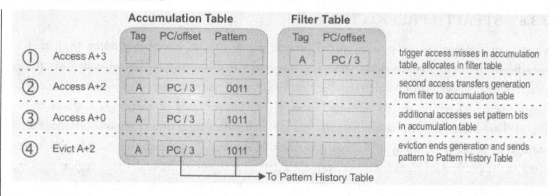

Figure 3.7: Structures for storage-efficient training in spatial memory streaming. From [82].

Whereas the active generation table is effective in reducing the storage requirements for training the SMS prefetcher, the required pattern history table size remains large, on the order of 64KB, for maximum effectiveness. Hence, follow-on work by Burcea and co-authors proposed virtualizing the pattern history table. Instead of a large dedicated table, the virtualized approach stores prefetcher meta-data in the last level cache, using a small, dedicated meta-data cache to accelerate access [89].

3.3.8   SPATIO-TEMPORAL MEMORY STREAMING

While effective for the targeted memory access patterns, spatial prefetchers are generally ineffective for pointer-based data structures with arbitrary memory layouts, and have shown limited effectiveness for some workloads with many pointer-chasing access patterns, such as on-line transaction processing [90]. Conversely, address correlating prefetchers are ineffective when data structure traversals do not repeat frequently. For example, memory copy and scan access patterns are easily recognized by stride and spatial prefetchers, yet the scans may recur too infrequently for address correlating prefetchers to capture [55]. The latest prefetcher proposals hence seek to integrate both address correlating and spatial prefetching into a single unified design.

Spatio-temporal memory streaming [91] integrates TMS (see Section 3.2.9) and SMS (see Section 3.3.7). A straight-forward integration of the two mechanisms might use TMS to supply a predicted stream of future trigger accesses (base address and PC), which are then fed to SMS to predict the remaining prefetch addresses within each region. While this simple approach is effective in predicting future accesses, it overwhelms the memory system, providing enormous bursts of prefetch requests, as the natural ordering and throttling of the individual mechanisms are lost.

To address this deficiency, spatio-temporal memory streaming instead seeks to reconstruct the total order of prefetch requests from the addresses individually predicted by the spatial and temporal mechanisms, respectively. SMS is enhanced to maintain miss ordering information by encoding spatial patterns as ordered streams of offsets. Although less compact than the bit vectors used in SFP, SPP, and conventional SMS, the offset streams maintain ordering information required to properly interleave spatial and temporal streams. As in the naïve integration, the TMS mechanism provides a sequence of triggers—base address and PC pairs—for SMS lookups. However, entries in both spatial and temporal streams are further augmented with a *delta*, which indicates the number of prefetch addresses from some other stream that interleave between two consecutive addresses in a particular spatial or temporal stream. During prefetching, these deltas are used to reconstruct the total miss order by entering prefetch addresses from a temporal stream into a reconstruction buffer while leaving empty space to be filled in by addresses supplied from subsequent spatial streams. Details of the reconstruction mechanism appear in [91].

3.4   EXECUTION-BASED PREFETCHING

The final category of data prefetcher we briefly address relies neither on repetition in miss sequences nor in data layouts; rather *execution-based prefetchers* seek to explore the program's instruction sequence ahead of instruction execution and retirement to discover address calculations and dereference pointers. The key objective of such prefetchers is to run faster than instruction execution itself, to get ahead of the processor core, while still using the actual address calculation algorithm to identify prefetch candidates. As such, these mechanisms do not rely on repetition at all. Instead, they rely on mechanisms that either summarize address calculation while omitting other aspects of the computation, guess at values directly, or leverage stall cycles and idle processor resources to explore ahead of instruction retirement.

3.4.1   ALGORITHM SUMMARIZATION

Several prefetching techniques summarize the instruction sequence that traverses a data structure, such that the traversal pattern can be executed faster than the main thread to prefetch data structure elements. Roth and co-authors [44, 45] propose a mechanism that summarizes traversals entirely in hardware by identifying pointer loads (load instructions that dereference a pointer) and the dependent chain of instructions that connect them. These dependence relationships are then encoded by hardware into a compact state machine, which can iterate through the sequence of dependent loads faster than instruction execution. Annavaram, Patel, and Davidson propose a general mechanism for extracting program dependence graphs—a subset of instructions that lead to missing loads—in hardware and then executing these graphs in dedicated precomputation engines [92].

3.4.2   HELPER-THREAD AND HELPER-CORE APPROACHES

Thread-based data prefetching techniques [93, 94, 95, 96, 97, 98, 99, 100, 101] use idle contexts on a multithreaded or multicore processor to run helper threads that overlap misses with speculative execution. Individual techniques vary in whether they are automatic or require compiler/software support, whether they rely on simultaneous multithreading hardware and specific thread coordination mechanisms, whether they rely on additional cores, and whether they require additional mechanisms to insert blocks into remote caches. In nearly all cases, these techniques repurpose spare execution contexts to execute the prefetching code. However, the spare resources the helper threads require (e.g., idle cores or thread contexts; fetch and execution bandwidth) may not be available when the processor executes an application exhibiting high thread-level parallelism. The benefit of these techniques must be weighed against scaling up the number of application threads.

3.4.3   RUN-AHEAD EXECUTION

Run-ahead execution uses the execution resources of a core that would otherwise be stalled on a long-latency event (e.g., off-chip cache miss) to explore ahead of the stalled execution in an effort to discover additional load misses and warm branch predictors. The idea in run-ahead is to capture a snapshot of execution state when the core would otherwise stall, then proceed past stalled instructions to continue to fetch and execute the predicted instruction stream. Instructions that are data-dependent on an incomplete instruction are not executed (e.g., a poison token is propagated through the register renaming mechanism). When the long-latency event resolves (e.g., the original miss returns), execution state is recovered from the snapshot and the original execution continues, re-crossing the instructions that were explored during run-ahead. The primary benefit of this scheme is the prefetching effect for long-latency loads. Run-ahead was originally proposed in the context of in-order cores by Dundas and Mudge [102]. Mutlu and co-authors explore efficient implementations in the context of out-of-order processors [103, 104, 105, 106]. More recently, authors have explored non-blocking pipeline microarchitectures that speculate past long-latency loads without discarding speculative execution results when the loads return, instead re-executing only the dependent instructions [107, 108].

3.4.4   CONTEXT RESTORATION

With the increasing prevalence of virtualization and multiplexing of multiple applications on a single server in cloud environments, an important special-case source of instruction cache misses arises due to context switches. Each time the operating system or hypervisor switches among multiplexed applications or virtual machines, the cache state of one application is overwritten by the other. These misses are particularly damaging when multiplexing a latency-sensitive serving application with background or batch tasks—a common practice to attempt to get value from otherwise-idle cycles when the serving application is blocked waiting for user requests.

Context restoration prefetchers [109, 110, 111] seek to capture cache contents at the time of a context switch (or as blocks are replaced after a context switch) and then restore these blocks to the cache when the context resumes execution. Unlike other prefetching techniques, these mechanisms do not need to make a prediction as to which addresses to prefetch, the objective is simply to restore what was previously in the cache. Instead, the design challenges center on associating cache state with a particular execution context, managing the timeliness of the prefetch and efficiently recording/storing the prefetcher metadata.

3.4.5   COMPUTATION SPREADING

An orthogonal approach to improve cache locality that bears some similarity to execution-based prefeching is *computation spreading* [112, 113]. Computation spreading uses thread migration to split the execution of a large computation across multiple cores, grouping similar execution fragments on each core. Repeated execution of similar fragments has the effect of specializing core-private instruction and data caches (as well as other structures, such as branch predictors) for each kind of fragment, improving locality. It has been demonstrated for separating application and OS code fragments [112] and for phases of online transaction processing threads [113].

3.5   PREFETCH MODULATION AND CONTROL

A variety of authors study prefetch modulation and control techniques and the interaction of prefetching with other aspects of the memory system, such as cache replacement policy. While most studies are performed in the context of strided stream prefetchers, these mechanisms are often applicable regardless of the particular algorithm used to generate candidate addresses for prefetch.

Srinath and co-authors propose hardware for tracking the cache pollution that can be caused by aggressive prefetch and dynamically reducing prefetcher aggressiveness when accuracy and timeliness are poor or when pollution substantially increases demand miss rates [114]. Ebrahimi and co-authors examine mechanisms to coordinate the aggressiveness of multiple per-core prefetchers in multi-core systems to address bandwidth contention [115]. Lee and colleagues examine the interaction of prefetchers and DRAM bank contention, and propose re-ordering prefetch requests to try to improve DRAM bank-level parallelism, improving prefetch throughput [116]. Hur and Lin propose mechanisms to dynamically detect the end of streams, reducing cache pollution and wasted bandwidth due to overfetching from short streams [33]. Lin et al. propose inserting prefetched blocks at the least-recently-used position in associative caches to manage pollution and modulating prefetcher issue rate [117]. Wu and colleagues build on this theme and propose several dynamic methods for choosing where in the cache replacement stack a prefetched block should be inserted [118]. Finally, Verma, Koppelman, and Peng provide mechanisms for software control of prefetch aggressiveness [119].

3.6 SOFTWARE APPROACHES

Researchers have proposed a wide variety of other approaches for compiler- or programmer-inserted prefetch instructions (e.g., [41, 43, 120, 121, 122, 123]) or data forwarding operations [124, 125]. More recently, researchers have developed architectures [126, 127] and programming languages [128, 129] that provide constructs for directly expressing and manipulating data streams. However, these studies have focused primarily on multimedia applications, where application inputs map naturally to data sequences. It remains unclear how to exploit these advances in other contexts, such as commercial server software. A thorough exploration of software and compiler data prefetching techniques is beyond the scope of this synthesis lecture.

Table 3.1: Summary of data prefetching techniques

Technique	Line of Attack	Lookahead	Accuracy	Cost/ Complexity
Stride/Stream	Predicts cache miss strides using PCs or addresses	A few cache blocks	High for strides	< 1 KB
Simple address correlating	Correlates one address to one or more distinct cache miss addresses	A single cache block	> 30%	MBs
Linked-data correlating	Custom FSM that uses load-to-use dependencies in loops to fetch pointers	A few cache blocks	High for specific data structures	< 1 KB
Dead-block correlating	Correlates up to two addresses and control flow to a subsequent address	Time from death event to replacement	> 50%	MBs
Temporal streaming	Correlates one or more addresses to a stream of cache misses	Many cache blocks	> 50%	Off chip
Chained streaming	Chaining multiple address streams together using control flow or hierarchical address prediction	Many cache blocks	> 50%	Off chip
Irregular stream buffer	Adds a layer of address indirection to convert temporal to spatial correlation	Many cache blocks	> 50%	< 32 KB

Delta correlation	Records and correlates sequences of deltas between miss addresses	A few cache blocks	> 30%	< 32 KB
Spatial streaming	Correlates a PC with an arbitrary sequence of cache miss address deltas	A few cache blocks	> 50%	< 64 KB
Execution based	Helper threads or support for speculative execution	Limited by execution bandwidth	varies	--

CHAPTER 4

Concluding Remarks

Hardware prefetching has been a subject of academic research and industrial development for over 40 years. Nevertheless, because of the scaling trends that continue to widen the gap between processor performance and memory access latency, the importance of hardware prefetching and the need to hide memory system latency has only grown—further innovation remains critical.

In this primer, we have surveyed the myriad of prefetching techniques that have been developed and highlighted the principle program behaviors on which these techniques are based. We hope this book serves as an introduction to the field, as an overview of the vast literature on hardware prefetching, and as a catalyst to spur new research efforts.

A number of challenges remain to be addressed in future work. Instruction fetch remains a fundamental bottleneck especially in servers with complex software stacks and ever-growing on-chip instruction working sets. Although instruction footprints can often fit entirely on chip in large last-level caches, cycle time constraints place severe limits on the capacity of L1 instruction caches, and access latency to larger caches remains exposed. Advanced proposals for temporal instruction streaming have in recent years achieved phenomenal accuracies and coverage (> 99.5%) even in the presence of complex software stacks. The key to a wider adoption of these proposals are techniques to reduce on-chip meta-data storage to practical levels.

Another key challenge to instruction prefetching is due to developments in programming languages and software engineering often complicating or even thwarting the techniques we have discussed. Object-oriented programming practices, dynamic dispatch, and managed runtimes all lead to an increase in the use of frequent, short function calls, indirection through function pointers, register-indirect branches and multi-way control transfers. Dynamic code generation/optimization, interpreted languages, and just-in-time compilation lead to environments where the control structure of a program may be obscured and instruction addresses change meaning over time. Virtualization and operating system layering similarly complicate and obscure control flow through frequent virtual-machine exits and traps to emulate privileged functionality.

On the hardware front, processors are increasingly supporting multiple concurrent hardware threads that must share already over-subscribed instruction cache capacity. Prefetchers must be enhanced to share limited capacity and bandwidth among threads, disambiguate instruction streams issuing from each thread, and consider the interaction of prefetching policies and thread prioritization/fetch policies.

A key remaining challenge for data prefetching is low accuracy and coverage across a broad spectrum of workloads. While the emergence of data-intensive workloads and large-scale in-memory data services is placing ever-growing demands on the need for effective data prefetching, the increase in memory capacity is dwarfing even the most advanced history-based prefetching techniques we cover in this primer in terms of diminishing repetitive history patterns and prohibitive meta-data storage requirements. Future advances in data prefetching are required to capture repetitive access patterns with lower meta-data storage requirement and with a higher accuracy.

Prefetching techniques are beginning to emerge for graphics processing units and other forms of specialized accelerators, which may have markedly different code and data access patterns than conventional processors. In the case of graphics processors, memory access stalls and thread/warp scheduling interact in complex ways, creating new opportunities for synergistic designs.

A fundamental challenge that has emerged in the past decade is that power has become a first-class constraint due to a slowdown in Dennard Scaling [130, 131] and leveling off of supply voltages. On the one hand, prefetchers eliminate stalls, which can lead to energy efficiency gains due to more efficient use of hardware resources. On the other hand, most prefetchers require auxiliary hardware structures, which require energy. Moreover, prefetchers often fetch incorrect blocks, which can waste substantial energy. Indeed, many of the simpler (but widely deployed) designs are wildly inaccurate; over half the blocks they retrieve may never be accessed. Advances in prefetching must target energy efficiency as a first-class constraint in conjunction with other key metrics such as accuracy and coverage in evaluating the effectiveness of a prefetcher design.

Bibliography

[1] W. A. Wulf and S. A. McKee. "Hitting the Memory Wall: Implications of the Obvious." *ACM SIGARCH Computer Architecture News*, v. 23 no. 1, 1995. DOI: 10.1145/216585.216588. xiii

[2] D. Lustig, A. Bhattacharjee, and M. Martonosi. "TLB Improvements for Chip Multiprocessors: Inter-Core Cooperative Prefetchers and Shared Last-Level TLBs." *ACM Transactions on Architecture and Code Optimization*, v. 10, no. 1, 2013. DOI: 10.1145/2445572.2445574. xiii

[3] J. Hennessy and D. A. Patterson. *Computer Architecture: A Quantitative Approach*, 4th ed. DOI: 10.1.1.115.1881. 1

[4] B. Jacob. "The Memory System: You Can't Avoid It, You Can't Ignore It, You Can't Fake It." *Synthesis Lectures on Computer Architecture*, v. 4, no. 1, 2009. DOI: 10.2200/S00201ED1V01Y200907CAC007. 2

[5] A. J. Smith. "Sequential Program Prefetching in Memory Hierarchies." *Computer*, v. 11, no. 12, 1978. DOI: 10.1109/C-M.1978.218016. 7, 15

[6] D. W. Anderson, F. J. Sparacio, and R. M. Tomasulo. "The IBM System/360 Model 91: Machine Philosophy and Instruction-Handling." *IBM Journal of Research and Development*, v. 11 no. 1, 1967. DOI: 10.1147/rd.111.0008. 8

[7] M. Ferdman, T. F. Wenisch, A. Ailamaki, B. Falsafi, A. Moshovos. "Temporal Instruction Fetch Streaming." In *Proc. of the 41st Annual ACM/IEEE International Symposium on Microarchitecture*, 2008. DOI: 10.1109/MICRO.2008.4771774. 8, 10, 12

[8] P. Ranganathan, K. Gharachorloo, S. V. Adve, and L. A. Barroso. "Performance of Database Workloads on Shared-Memory Systems With Out-Of-Order Processors." In *Proc. of the 8th International Conference on Architectural Support for Programming Languages and Operating Systems*, 1998. DOI: 10.1145/291069.291067. 8

[9] A. Ramirez, O. J. Santana, J. L. Larriba-Pey and M. Valero. "Fetching Instruction Streams." In *Proc. of the 35th Annual ACM/IEEE International Symposium on Microarchitecture*, 2002. 8

[10] O. J. Santana, A. Ramirez, and M. Valero. "Enlarging Instruction Streams." *IEEE Transactions on Computers*, v. 56, no. 10, 2007. DOI: 10.1109/TC.2007.70742. 8, 11

[11]   I-C. K. Chen, C-C. Lee, and T. N. Mudge. "Instruction Prefetching Using Branch Prediction Information." In *Proc. of the IEEE International Conference on Computer Design*, 1997. DOI: 10.1109/ICCD.1997.628926. 9

[12]   G. Reinman, B. Calder, and T. Austin. "Fetch Directed Instruction Prefetching." In *Proc. of the 32nd Annual ACM/IEEE International Symposium on Microarchitecture*, 1999. 9

[13]   A. V. Veidenbaum, Q. Zhao, and A. Shameer. "Non-Sequential Instruction Cache Prefetching for Multiple–Issue Processors." *International Journal of High Speed Computing*, v. 10, no. 1, 1999. DOI: 10.1142/S0129053399000065. 9

[14]   R. Panda, P. V. Gratz, and D. A. Jiménez. "B-Fetch: Branch Prediction Directed Prefetching for In-Order Processors." In *Proc. of the 18th International Symposium on High-Performance Computer Architecture*, 2012. DOI: 10.1109/L-CA.2011.33. 9

[15]   T. Sherwood, S. Sair, and B. Calder. "Predictor-Directed Stream Buffers." In *Proc. of the 33rd Annual ACM/IEEE International Symposium on Microarchitecture*, 2000. DOI: 10.1145/360128.360135. 9

[16]   J. Pierce, and T. N. Mudge. "Wrong-Path Instruction Prefetching." In *Proc. of the 29th Annual ACM/IEEE International Symposium on Microarchitecture*, 1996. 11

[17]   V. Srinivasan, E. S. Davidson, G. S. Tyson, M. J. Charney, and T. R. Puzak. "Branch History Guided Instruction Prefetching." In *Proc. of the 7th International Symposium on High-Performance Computer Architecture*, 2001. DOI: 10.1109/HPCA.2001.903271. 11

[18]   Y. Zhang, S. Haga, and R. Barua. "Execution History Guided Instruction Prefetching." In *Proc. of the 16th Annual International Conference on Supercomputing*, 2002. DOI: 10.1145/514191.514220. 11

[19]   Q. Jacobson, E. Rotenberg, and J. E. Smith. "Path-Based Next Trace Prediction." In *Proc. of the 30th Annual ACM/IEEE International Symposium on Microarchitecture*, 1997. DOI: 10.1109/MICRO.1997.645793. 11

[20]   M. Annavaram, J. M. Patel, and E. S. Davidson. "Call Graph Prefetching for Database Applications." *ACM Transactions on Computer Systems*, v. 21, no. 4, 2003. DOI: 10.1145/945506.945509. 11

[21]   L. Spracklen, Y. Chou, and S. G. Abraham. "Effective Instruction Prefetching in Chip Multiprocessors for Modern Commercial Applications." In *Proc. of the 11th International Symposium on High-Performance Computer Architecture*, 2005. DOI: 10.1109/HPCA.2005.13. 11

[22]   T. M. Aamodt, P. Chow, P. Hammarlund, H. Wang, and J. P. Shen. "Hardware Support for Prescient Instruction Prefetch." *Proc. of the 10th International Symposium on High-Performance Computer Architecture*, 2004. DOI: 10.1109/HPCA.2004.10028. 12

[23]   C-K. Luk, T. C. Mowry. "Cooperative Prefetching: Compiler and Hardware Support for Effective Instruction Prefetching In Modern Processors." In *Proc. of the 31st annual ACM/IEEE International Symposium on Microarchitecture*, 1998. DOI: 10.1109/MICRO.1998.742780. 12

[24]   K. Sundaramoorthy, Z. Purser, and E. Rotenberg. "Slipstream Processors: Improving Both Performance and Fault Tolerance." In *Proc. of the 9th International Conference on Architectural Support for Programming Languages and Operating Systems*, 2000. DOI: 10.1145/356989.357013. 12

[25]   C. Zilles and G. Sohi. "Execution-Based Prediction Using Speculative Slices." In *Proc. of the 28th Annual International Symposium on Computer Architecture*, 2001. DOI: 10.1145/379240.379246. 12

[26]   A. Kolli, A. Saidi, and T. F. Wenisch. "RDIP: Return-Address-Stack Directed Instruction Prefetching." In *Proc. of the 46th Annual IEEE/ACM International Symposium on Microarchitecture*, 2013. DOI: 10.1145/2540708.2540731. 13

[27]   M. Ferdman, C. Kaynak, and B. Falsafi. "Proactive Instruction Fetch." In *Proc. of the 44th Annual IEEE/ACM International Symposium on Microarchitecture*, 2011. DOI: 10.1145/2155620.2155638. 14

[28]   C. Kaynak, B. Grot, and B. Falsafi. "Shift: Shared History Instruction Fetch for Lean-Core Server Processors." In *Proc. of the 46th Annual IEEE/ACM International Symposium on Microarchitecture*, 2013. DOI: 10.1145/2540708.2540732. 14

[29]   J.-L. Baer and T.-F. Chen. "An Effective On-Chip Preloading Scheme to Reduce Data Access Penalty." In *Proc. of Supercomputing*, 1991. DOI: 10.1145/125826.125932. 15, 16

[30]   F. Dahlgren and P. Stenstrom. "Effectiveness of Hardware-Based Stride and Sequential Prefetching in Shared-Memory Multiprocessors." In *Proc. of the 1st IEEE Symposium on High-Performance Computer Architecture*, 1995. DOI: 10.1109/HPCA.1995.386554. 16

[31]   Y. Ishii, M. Inaba and K. Hiraki. "Access Map Pattern Matching for High Performance Data Cache Prefetch." *Journal of Instruction-Level Parallelism*, v. 13, 2011. 16, 28

[32]   S. Sair, T. Sherwood, and B. Calder. "A Decoupled Predictor-Directed Stream Prefetching Architecture." *IEEE Transactions on Computers*, v. 52, no. 3, 2003. DOI: 10.1109/TC.2003.1183943. 16

[33]   I. Hur and C. Lin. "Memory Prefetching Using Adaptive Stream Detection." In *Proc. of the 39th Annual ACM/IEEE International Symposium on Microarchitecture*, 2006. DOI: 10.1109/MICRO.2006.32. 16, 35

[34]   N. P. Jouppi. "Improving Direct-Mapped Cache Performance by the Addition of a Small Fully-Associative Cache and Prefetch Buffers." In *Proc. of the 17th Annual International Symposium on Computer Architecture*, 1990. DOI: 10.1145/325164.325162. 16, 24

[35]   S. Palacharla and R. E. Kessler. "Evaluating Stream Buffers As a Secondary Cache Placement." In *Proc. of the 21st Annual International Symposium on Computer Architecture*, 1994. 17

[36]   C. Zhang and S. A. McKee. "Hardware-Only Stream Prefetching and Dynamic Access Ordering." In *Proc. of the 14th Annual International Conference on Supercomputing*, 2000. DOI: 10.1145/335231.335247. 17

[37]   S. Iacobovici, L. Spracklen, S. Kadambi, Y. Chou and S. G. Abraham. "Effective Stream-Based and Execution-Based Data Prefetching." In *Proc. of the 18th Annual International Conference on Supercomputing*, 2004. DOI: 10.1145/1006209.1006211. 17

[38]   J-L. Baer, J-L., and G. R. Sager. "Dynamic Improvement of Locality in Virtual Memory Systems." *IEEE Transactions on Software Engineering*, v. 1, 1976. DOI: 10.1109/TSE.1976.233801. 17

[39]   M. J. Charney and A. P. Reeves. "Generalized Correlation-Based Hardware Prefetching." *Technical Report EE-CEG-95-1*, School of Electrical Engineering, Cornell University, Feb. 1995. 17

[40]   M. J. Charney. *Correlation-Based Hardware Prefetching*, 1996. Ph.D. diss., Cornell University, 1996. 17

[41]   T. M. Chilimbi and M. Hirzel. "Dynamic Hot Data Stream Prefetching for General-Purpose Programs." In *Proc. of the Conference on Programming Language Design and Implementation*, 2002. DOI: 10.1145/512529.512554. 17, 19, 20, 24, 36

[42]   T. F. Wenisch, S. Somogyi, N. Hardavellas, J. Kim, A. Ailamaki, and B. Falsafi. "Temporal Streaming of Shared Memory." In *Proc. of the 32nd Annual International Symposium on Computer Architecture*, June 2005. DOI: 10.1109/ISCA.2005.50. 17, 20, 22

[43]   C.-K. Luk and T. C. Mowry. "Compiler Based Prefetching for Recursive Data Structures." In *Proc. of the 7th International Conference on Architectural Support for Programming Languages and Operating Systems*, 1996. DOI: 10.1145/237090.237190. 17, 36

[44]   A. Roth, A. Moshovos, and G. S. Sohi. "Dependence Based Prefetching for Linked Data Structures." In *Proc. of the 8th International Conference on Architectural Support for Programming Languages and Operating Systems*, 1998. DOI: 10.1145/291069.291034. 17, 33

[45]   A. Roth and G. S. Sohi. "Effective Jump Pointer Prefetching for Linked Data Structures." In *Proc. of the 26th Annual International Symposium on Computer Architecture*, 1999. DOI: 10.1109/ISCA.1999.765944. 17, 33

[46]   J. Collins, S. Sair, B. Calder, and D. M. Tullsen. "Pointer Cache Assisted Prefetching." In *Proc. of the 35th Annual ACM/IEEE International Symposium on Microarchitecture*, 2002. DOI: 10.1109/MICRO.2002.1176239. 17

[47]   R. Cooksey, S. Jourdan, and D. Grunwald. "A Stateless, Content-Directed Data Prefetching Mechanism." In *Proc. of the 10th International Conference on Architectural Support for Programming Languages and Operating Systems*, 2002. DOI: 10.1145/605397.605427. 18

[48]   E. Ebrahimi, O. Mutlu, and Y. N. Patt. "Techniques for Bandwidth-Efficient Prefetching of Linked Data Structures in Hybrid Prefetching Systems." In *Proc. of the 15th International Symposium on High Performance Computer Architecture*, 2009. DOI: 10.1109/HPCA.2009.4798232. 18

[49]   D. Joseph and D. Grunwald. "Prefetching Using Markov Predictors." In *Proc. of the 24th Annual International Symposium on Computer Architecture*, 1997. DOI: 10.1145/264107.264207. 18, 19

[50]   D. Joseph and D. Grunwald. "Prefetching Using Markov Predictors." *IEEE Transactions on Computers*, v. 48 no. 2, 1999. DOI: 10.1109/12.752653. 18

[51]   A.-C. Lai, C. Fide, and B. Falsafi. "Dead-Block Prediction and Dead-Block Correlating Prefetchers." In *Proc. of the 28th Annual International Symposium on Computer Architecture*, 2001. DOI: 10.1145/379240.379259. 19, 20, 21

[52]   Y. Solihin, J. Lee, and J. Torrellas. "Using a User-Level Memory Thread for Correlation Prefetching." In *Proc. of the 29th Annual International Symposium on Computer Architecture*, May 2002. DOI: 10.1109/ISCA.2002.1003576. 20, 22

[53]   Y. Solihin, J. Lee, and J. Torrellas. "Correlation Prefetching with a User-Level Memory Thread." *IEEE Transactions on Parallel and Distributed Systems*, v. 14, no. 6, 2003. DOI: 10.1109/TPDS.2003.1206504. 20

[54]   M. Ferdman and B. Falsafi. "Last-Touch Correlated Data Streaming." In *IEEE International Symposium on Performance Analysis of Systems and Software*, 2007. DOI: 10.1109/ISPASS.2007.363741. 20, 21, 22

[55]   T. F. Wenisch, M. Ferdman, A. Ailamaki, B. Falsafi, and A. Moshovos. "Temporal Streams in Commercial Server Applications." In *Proc. of the IEEE International Symposium on Workload Characterization*, 2008. DOI: 10.1109/IISWC.2008.4636095. 20, 22, 32

[56]   Y. Chou, B. Fahs, and S. Abraham. "Microarchitecture Optimizations for Exploiting Memory-Level Parallelism." In *Proc. of the 31st Annual International Symposium on Computer Architecture*, 2004. DOI: 10.1145/1028176.1006708. 20

[57]   Y. Chou. "Low-Cost Epoch-Based Correlation Prefetching for Commercial Applications." In *Proc. of the 40th Annual ACM/IEEE International Symposium on Microarchitecture*, 2007. DOI: 10.1109/MICRO.2007.39. 20

[58]   N. Kohout, S. Choi, D. Kim, and D. Yeung. "Multi-Chain Prefetching: Effective Exploitation of Inter-Chain Memory Parallelism for Pointer-Chasing Codes." In *Proc. of the International Conference on Parallel Architectures and Compilation Techniques*, 2001. DOI: 10.1109/PACT.2001.953307. 24

[59]   P. Díaz and M. Cintra. "Stream Chaining: Exploiting Multiple Levels of Correlation in Data Prefetching." In *Proc. of the 36th Annual International Symposium on Computer Architecture*, 2009. DOI: 10.1145/1555754.1555767. 24

[60]   A-C. Lai, and B. Falsafi. "Selective, Accurate, and Timely Self-Invalidation Using Last-Touch Prediction." In *Proc. of the 27th Annual International Symposium on Computer Architecture*, 2000. DOI: 10.1145/339647.339669. 20, 28

[61]   Z. Hu, S. Kaxiras, and M. Martonosi. "Timekeeping in the Memory System: Predicting and Optimizing Memory Behavior." In *Proc. of the 29th Annual International Symposium on Computer Architecture*, 2002. DOI: 10.1109/ISCA.2002.1003579. 20, 21

[62]   H. Liu, M. Ferdman, J. Huh, and D. Burger. "Cache Bursts: A New Approach for Eliminating Dead Blocks and Increasing Cache Efficiency." In *Proc. of the 41st Annual ACM/IEEE International Symposium on Microarchitecture*, 2008. DOI: 10.1109/MICRO.2008.4771793. 20, 21

[63]   T. R. Puzak. *Analysis of Cache Replacement-Algorithms*, 1985. Ph.D. diss., Univ. Massachusetts, Amherst, 1985. 20

[64]   A. Mendelson, D. Thiebaut, and D. K. Pradhan. "Modeling Live and Dead Lines in Cache Memory Systems." *IEEE Transactions on Computers*, v. 42, no. 1. DOI: 10.1109/12.192209. 20

[65]   D. A. Wood, M. D. Hill, and R. E. Kessler. "A Model for Estimating Trace-Sample Miss Ratios." In *Proc. of the 1991 ACM SIGMETRICS Conference on Measurement and Modeling of Computer Systems*, 1991. DOI: 10.1145/107971.107981. 20

[66] Z. Hu, M. Martonosi, and S. Kaxiras. "TCP: Tag Correlating Prefetchers." In *Proc. of the 9th IEEE Symposium on High-Performance Computer Architecture*, 2003. DOI: 10.1109/HPCA.2003.1183549. 21

[67] T. F. Wenisch, M. Ferdman, A. Ailamaki, B. Falsafi, A. Moshovos. "Practical Off-Chip Meta-Data for Temporal Memory Streaming." In *Proc. of the 15th International Symposium on High Performance Computer Architecture*, 2009. DOI: 10.1109/HPCA.2009.4798239. 22, 25

[68] K. J. Nesbit and J. E. Smith. "Data Cache Prefetching Using a Global History Buffer." In *Proc. of the 10th IEEE Symposium on High-Performance Computer Architecture*, 2004. DOI: 10.1109/HPCA.2004.10030. 22, 23, 24, 26, 27, 28

[69] K. J. Nesbit, A. S. Dhodapkar, and J. E. Smith. "AC/DC: An Adaptive Data Cache Prefetcher." In *Proc. of the 13th International Conference on Parallel Architectures and Compilation Techniques*, 2004. DOI: 10.1109/PACT.2004.1342548. 23, 24, 27

[70] T. F. Wenisch, M. Ferdman, A. Ailamaki, B. Falsafi and A. Moshovos. "Making Address-Correlated Prefetching Practical." *IEEE Micro*, v. 30, no. 1, 2010. DOI: 10.1109/MM.2010.21. 25

[71] I. Chung, C. Kim, H.-F. Wen, and G. Cong. "Application Data Prefetching on the IBM Blue Gene/Q Supercomputer." In *International Conference on High Performance Computing, Networking, Storage and Analysis*, 2012. DOI: 10.1109/SC.2012.19. 25

[72] A. Jain and C. Lin. "Linearizing Irregular Memory Accesses for Improved Correlated Prefetching." In *Proc. of the 46th Annual ACM/IEEE International Symposium on Microarchitecture*, 2013. DOI: 10.1145/2540708.2540730. 22, 25

[73] G. B. Kandiraju and A. Sivasubramaniam. "Going the Distance for Tlb Prefetching: An Application-Driven Study." In *Proc. of the 29th Annual International Symposium on Computer Architecture*, 2002. DOI: 10.1109/ISCA.2002.1003578. 27

[74] M. Grannaes, M. Jahre, and L. Natvig. "Storage Efficient Hardware Prefetching Using Delta Correlating Prediction Tables." *Journal of Instruction-Level Parallelism*, v. 13, 2011. 28

[75] M. Dimitrov and H. Zhou. "Combining Local and Global History for High Performance Data Prefetching." *Journal of Instruction-Level Parallelism*, v. 13, 2011. 28

[76] G. Liu, Z. Huang, J-K. Peir, X. Shi, and L. Peng. "Enhancements for Accurate and Timely Streaming Prefetcher." *Journal of Instruction-Level Parallelism*, v. 13, 2011. 28

[77]   L. M. Ramos, J. L. Briz, P. E. Ibáñez, and V. Viñals. "Multi-Level Adaptive Prefetching Based on Performance Gradient Tracking." *Journal of Instruction-Level Parallelism*, v. 13, 2011. 28

[78]   A. Sharif and H-H. Lee. "Data Prefetching by Exploiting Global and Local Access Patterns." *Journal of Instruction-Level Parallelism*, v. 13, 2011. 28

[79]   S. S. Mukherjee and M. D. Hill. "Using Prediction to Accelerate Coherence Protocols." In *Proc. of the 25th Annual International Symposium on Computer Architecture*, 1998. DOI: 10.1109/ISCA.1998.694773. 28

[80]   S. Kaxiras, J. R. Goodman. "Improving CC-NUMA Performance Using Instruction-Based Prediction." In *Proc. of the 5th International Symposium on High-Performance Computer Architecture*, 1999. DOI: 10.1109/HPCA.1999.744359. 28

[81]   S. Kumar and C. Wilkerson. "Exploiting Spatial Locality in Data Caches Using Spatial Footprints." In *Proc. of the 25th Annual International Symposium on Computer Architecture*, 1998. DOI: 10.1145/279358.279404. 28, 30

[82]   S. Somogyi, T. F. Wenisch, A. Ailamaki, B. Falsafi, and A. Moshovos. "Spatial Memory Streaming." In *Proc. of the 33rd Annual International Symposium on Computer Architecture*, 2006. DOI: 10.1109/ISCA.2006.38. 29, 31, 32

[83]   C. F. Chen, S.-H. Yang, B. Falsafi, and A. Moshovos. "Accurate and Complexity-Effective Spatial Pattern Prediction." In *Proc. of the 10th IEEE Symposium on High-Performance Computer Architecture*, Feb. 2004. DOI: 10.1109/HPCA.2004.10010. 29, 30

[84]   M. Ferdman, S. Somogyi, and B. Falsafi. "Spatial Memory Streaming with Rotated Patterns." *1st JILP Data Prefetching Championship*, 2009. 29

[85]   S. Somogyi, T. F. Wenisch, M. Ferdman, and B. Falsafi. "Spatial Memory Streaming." *Journal of Instruction-Level Parallelism*, v. 13, 2011. DOI: 10.1109/ISCA.2006.38. 29

[86]   A. Seznec. "Decoupled Sectored Caches: Conciliating Low Tag Implementation Cost and Low Miss Ratio." In *Proc. of the 21st Annual International Symposium on Computer Architecture*, 1994. DOI: 10.1145/191995.192072. 30

[87]   M. D. Powell, S-H. Yang, B. Falsafi, K. Roy, and T. N. Vijaykumar. "Gated-Vdd: A Circuit Technique to Reduce Leakage in Deep-Submicron Cache Memories." In *Proc. of the International Symposium on Low Power Electronics and Design*, 2000. DOI: 10.1145/344166.344526. 30

[88]   J. F. Cantin, M. H. Lipasti, and J. E. Smith. "Stealth Prefetching." In *Proc. of the 12th International Conference on Architectural Support for Programming Languages and Operating Systems*, 2006. DOI: 10.1145/1168857.1168892. 31

[89]   I. Burcea, S. Somogyi, A. Moshovos, and B. Falsafi. "Predictor Virtualization." In *Proc. of the 13th International Conference on Architectural Support for Programming Languages and Operating Systems*, 2008. DOI: 10.1145/1346281.1346301. 32

[90]   T. F. Wenisch. *Temporal memory streaming.* Ph.D. diss., Carnegie Mellon University, 2007. 32

[91]   S. Somogyi, T. F. Wenisch, A. Ailamaki, and B. Falsafi. "Spatio-Temporal Memory Streaming." In *Proc. of the 36th Annual International Symposium on Computer Architecture*, 2009. DOI: 10.1145/1555754.1555766. 32, 33

[92]   M. Annavaram, J. M. Patel, and E. S. Davidson. "Data Prefetching by Dependence Graph Precomputation." In *Proc. of the 28th Annual International Symposium on Computer Architecture*, 2001. DOI: 10.1109/ISCA.2001.937432. 33

[93]   J. D. Collins, H. Wang, D. M. Tullsen, C. Hughes, Y.-F. Lee, D. Lavery, and J. P. Shen. "Speculative Precomputation: Long-Range Prefetching of Delinquent Loads." In *Proc. of the 28th Annual International Symposium on Computer Architecture*, 2001. DOI: 10.1145/379240.379248. 33

[94]   J. D. Collins, D. M. Tullsen, H. Wang, and J. P. Shen. "Dynamic Speculative Precomputation." In *Proc. of the 34th Annual ACM/IEEE International Symposium on Microarchitecture*, 2001. 33

[95]   I. Ganusov and M. Burtscher. "Future execution: A Hardware Prefetching Technique for Chip Multiprocessors." In *Proc. of the 14th International Conference on Parallel Architectures and Compilation Techniques*, 2005. DOI: 10.1109/PACT.2005.23. 33

[96]   I. Ganusov and M. Burtscher. "Future Execution: A Prefetching Mechanism that Uses Multiple Cores to Speed Up Single Threads." *ACM Transactions on Architecture and Code Optimization*, v. 3, no. 4, 2006. DOI: 10.1145/1187976.1187979. 33

[97]   J. Lee, C. Jung, D. Lim, and Y. Solihin. "Prefetching with Helper Threads for Loosely Coupled Multiprocessor Systems." *IEEE Transactions on Parallel and Distributed Systems*, v. 20, no. 9, 2009. DOI: 10.1109/TPDS.2008.224. 33

[98]   W. Zhang, D. M. Tullsen, and B. Calder. "Accelerating and Adapting Precomputation Threads for Efficient Prefetching." In *Proc. of the 13th International Symposium on High Performance Computer Architecture*, 2007. DOI: 10.1109/HPCA.2007.346187. 33

[99]   R. S. Chappell, F. Tseng, A. Yoaz, and Y. N. Patt. "Microarchitectural Support for Precomputation Microthreads." In *Proc. of the 35th Annual ACM/IEEE International Symposium on Microarchitecture*, 2002. DOI: 10.1109/MICRO.2002.1176240. 33

[100]   M. Kamruzzaman, S. Swanson, and D. M. Tullsen. "Inter-Core Prefetching for Multicore Processors Using Migrating Helper Threads." In *Proc. of the 16th International Conference on Architectural Support for Programming Languages and Operating Systems*, 2011. DOI: 10.1145/1950365.1950411. 33

[101]   A. Roth and G. S. Sohi. "Speculative Data-Driven Multithreading." In *Proc. of the 7th International Symposium on High-Performance Computer Architecture*, 2001. DOI: 10.1109/HPCA.2001.903250. 33

[102]   J. Dundas and T. N. Mudge. "Improving Data Cache Performance by Pre-Executing Instructions under a Cache Miss." In *Proc. of the 11th Annual International Conference on Supercomputing*, 1997. DOI: 10.1145/263580.263597. 34

[103]   O. Mutlu, J. Stark, C. Wilkerson, and Y. N. Patt. "Runahead Execution: An Alternative to Very Large Instruction Windows for Out-Of-Order Processors." In *Proc. of the 9th International Symposium on High-Performance Computer Architecture*, 2003. DOI: 10.1109/HPCA.2003.1183532. 34

[104]   O. Mutlu, J. Stark, C. Wilkerson, and Y. N. Patt. "Runahead execution: An Effective Alternative to Large Instruction Windows." *IEEE Micro*, v. 23, no. 6, 2003. DOI: 10.1109/MM.2003.1261383. 34

[105]   O. Mutlu, H. Kim, and Y. N. Patt. "Techniques for Efficient Processing in Runahead Execution Engines." In *Proc. of the 32nd Annual International Symposium on Computer Architecture*, 2005. DOI: 10.1109/ISCA.2005.49. 34

[106]   O. Mutlu, H. Kim, and Y. N. Patt. "Efficient Runahead Execution: Power-Efficient Memory Latency Tolerance." *IEEE Micro*, v. 26, no. 1, 2006. DOI: 10.1109/MM.2006.10. 34

[107]   S. T. Srinivasan, R. Rajwar, H. Akkary, A. Gandhi, and M. Upton. "Continual Flow Pipelines." In *Proc. of the International Conference on Architectural Support for Programming Languages and Operating Systems*, 2004. DOI: 10.1145/1024393.1024407. 34

[108]   A. Hilton, S. Nagarakatte, and A. Roth. "iCFP: Tolerating All-Level Cache Misses in In-Order Processors." In *Proc. of the 15th International Symposium on High Performance Computer Architecture*, 2009. DOI: 10.1109/MM.2010.20. 34

[109]   H. Cui and S. Suleyman. "Extending Data Prefetching to Cope with Context Switch Misses." In *Proc. of the International Conference on Computer Design*, 2009. DOI: 10.1109/ICCD.2009.5413144. 34, 35

[110]   D. Daly and H. W. Cain. "Cache Restoration for Highly Partitioned Virtualized Systems." In *Proc. of the 18th Annual International Symposium on High Performance Computer Architecture*, 2012. DOI: 10.1109/HPCA.2012.6169029. 34

[111]   J. Zebchuk, H. W. Cain, X. Tong, V. Srinivasan and A. Moshovos. "RECAP: A Region-Based Cure for the Common Cold (Cache)." In *Proc. of the 19th Annual International Symposium on High Performance Computer Architecture*, 2013. DOI: 10.1145/2370816.2370887. 34

[112]   K. Chakraborty, P. M. Wells, and G. S. Sohi. "Computation Spreading: Employing Hardware Migration to Specialize CMP Cores On-The-Fly." In *Proc. of the 12th International conference on Architectural Support for Programming Languages and Operating Systems*, 2006. DOI: 10.1145/1168857.1168893. 35

[113]   I. Atta, P. Tozun, A. Ailamaki, and A. Moshovos. "SLICC: Self-Assembly of Instruction Cache Collectives for OLTP Workloads." In *Proc. of the 2012 45th Annual ACM/IEEE International Symposium on Microarchitecture*, 2012. DOI: 10.1109/MICRO.2012.26. 35

[114]   S. Srinath, O. Mutlu, H. Kim, and Y. N. Patt. "Feedback Directed Prefetching: Improving the Performance and Bandwidth-Efficiency of Hardware Prefetchers." In *Proc. of the 13th International Symposium on High Performance Computer Architecture*, 2007. DOI: 10.1109/HPCA.2007.346185.

[115]   E. Ebrahimi, O. Mutlu, C. J. Lee, Y. N. Patt. "Coordinated Control of Multiple Prefetchers in Multi-Core Systems." In *Proc. of the 42nd Annual ACM/IEEE International Symposium on Microarchitecture*, 2009. DOI: 10.1145/1669112.1669154. 35

[116]   C. J. Lee, V. Narasiman, O. Mutlu, Y. N. Patt, "Improving Memory Bank-Level Parallelism in the Presence of Prefetching." In *Proc. of the 42nd Annual ACM/IEEE International Symposium on Microarchitecture*, 2009. DOI: 10.1145/1669112.1669155. 35

[117]   W.-f. Lin, S. Reinhardt, and D. Burger. "Reducing DRAM Latencies with an Integrated Memory Hierarchy Design." In *Proc. of the 7th International Symposium on High Performance Computer Architecture*, 2001. DOI: 10.1109/HPCA.2001.903272. 35

[118]   C-J. Wu, A. Jaleel, M. Martonosi, S. Steely Jr, and J. Emer. "PACMan: Prefetch-Aware Cache Management for High Performance Caching." In *Proc. of the 44th Annual ACM/IEEE International Symposium on Microarchitecture*, 2011. DOI: 10.1145/2155620.2155672. 35

[119]   S. Verma, D. M. Koppelman, and L. Peng. "Efficient Prefetching with Hybrid Schemes and Use of Program Feedback to Adjust Prefetcher Aggressiveness." *Journal of Instruction-Level Parallelism*, v. 13, 2011. 35

[120]   S. Chen, A. Ailamaki, P. B. Gibbons, and T. C. Mowry. "Improving Hash Join Performance through Prefetching." In *Proc. of the 20th International Conference on Data Engineering*, 2004. DOI: 10.1109/ICDE.2004.1319989. 36

[121]   S. Chen, P. B. Gibbons, and T. C. Mowry. "Improving Index Performance through Prefetching." In *Proc. of the 20th Annual ACM International Conference on Management of Data*, 2001. DOI: 10.1145/375663.375688. 36

[122]   T. C. Mowry, M. S. Lam, and A. Gupta. "Design and Evaluation of a Compiler Algorithm for Prefetching." In *Proc. of the 5th International Conference on Architectural Support for Programming Languages and Operating Systems*, 1992. DOI: 10.1145/143365.143488. 36

[123]   Z. Wang, D. Burger, K. S. McKinley, S. K. Reinhardt, and C. C. Weems. "Guided Region Prefetching: A Cooperative Hardware/Software Approach." In *Proc. of the 30th Annual International Symposium on Computer Architecture*, 2003. DOI: 10.1145/859618.859663. 36

[124]   D. Koufaty, X. Chen, D. Poulsen, and J. Torrellas. "Data Forwarding in Scalable Shared-Memory Multiprocessors." In *Proc. of the 9th Annual International Conference on Supercomputing*, 1995. DOI: 10.1145/224538.224569. 36

[125]   C.-K. Luk and T. C. Mowry. "Memory Forwarding: Enabling Aggressive Layout Optimizations by Guaranteeing the Safety of Data Relocation." In *Proc. of the 26th Annual International Symposium on Computer Architecture*, 1999. DOI: 10.1145/300979.300987. 36

[126]   J. H. Ahn, W. J. Dally, B. Khailany, U. J. Kapasi, and A. Das. "Evaluating the Imagine Stream Architecture." In *Proc. of the 31st Annual International Symposium on Computer Architecture*, 2004. 36

[127]   W. J. Dally, F. Labonte, A. Das, P. Hanrahan, J.-H. Ahn, J. Gummaraju, M. Erez, N. Jayasena, I. Buck, T. J. Knight, and U. J. Kapasi. "Merrimac: Supercomputing with Streams." In *Proc. of Supercomputing*, 2003. DOI: 10.1145/1048935.1050187. 36

[128]   M. I. Gordon, W. Thies, M. Karczmarek, J. Lin, A. S. Meli, A. A. Lamb, C. Leger, J. Wong, H. Hoffmann, D. Maze, and S. Amarasinghe. "A Stream Compiler for Communication-Exposed Architectures." In *Proc. of the 10th International Conference on Architectural Support for Programming Languages and Operating Systems*, 2002. DOI: 10.1145/605397.605428. 36

[129]   J. Gummaraju and M. Rosenblum. "Stream Programming on General-Purpose Processors." In *Proc. of the 38th Annual ACM/IEEE International Symposium on Microarchitecture*, 2005. DOI: 10.1109/MICRO.2005.32. 36

[130]   H. Esmaeilzadeh, E. Blem, R. S. Amant, K. Sankaralingam, and D. Burger. "Dark Silicon and the End of Multicore Scaling." In *Proc. of the 38th Annual International Symposium on Computer Architecture*, 2011. DOI: 10.1145/2000064.2000108. 40

[131]   N. Hardavellas, M. Ferdman, B. Falsafi, and A. Ailamaki. "Toward Dark Silicon in Servers." In *IEEE Micro*, v. 31, no. 4, 2011. DOI: 10.1109/MM.2011.77. 40

Author Biographies

Babak Falsafi is a Professor in the School of Computer and Communication Sciences at EPFL, and the founding director of the EcoCloud research center, targeting future energy-efficient and environmentally friendly cloud technologies. He has made numerous contributions to computer system design and evaluation including: a scalable multiprocessor architecture that laid the foundation for the Sun (now Oracle) WildFire servers; snoop filters; temporal stream prefetchers that are incorporated into IBM BlueGene/P and BlueGene/Q; and computer system simulation sampling methodologies that have been in use by AMD and HP for research and product development. His most notable contribution has been to be first to show that, contrary to conventional wisdom, multiprocessor memory programming models—known as memory consistency models—prevalent in all modern systems are neither necessary nor sufficient to achieve high performance. He is a recipient of an NSF CAREER award, IBM Faculty Partnership Awards, and an Alfred P. Sloan Research Fellowship. He is a fellow of IEEE.

Thomas Wenisch is an Associate Professor of Computer Science and Engineering at the University of Michigan, specializing in computer architecture. His prior research includes memory streaming for commercial server applications, store-wait-free multiprocessor memory systems, memory disaggregation, and rigorous sampling-based performance evaluation methodologies. His ongoing work focuses on computational sprinting, memory persistency, data center architecture, energy-efficient server design, and accelerators for medical imaging. Wenisch received the NSF CAREER award in 2009 and the University of Michigan Henry Russel Award in 2013. He received his Ph.D. in Electrical and Computer Engineering from Carnegie Mellon University.